Two views of the *San Diego* soon after completion, when she was named *California*. Note the exquisite gilt carving on the bow, and the number of stars in the American flag. (Both courtesy of the San Francisco Maritime National Historical Park.)

U.S.S. SAN DIEGO

THE LAST ARMORED CRUISER

BY GARY GENTILE

GARY GENTILE PRODUCTIONS
P.O. Box 57137
Philadelphia, PA 19111
1989

Copyright 1989 by Gary Gentile

All rights reserved. Except for the use of brief quotations embodied in critical articles and reviews, this book may not be reproduced in part or in whole, in any manner (including mechanical, electronic, photographic, and photocopy means), transmitted in any form, or recorded by any data storage and/or retrieval system, without express written permission from the author. Address all queries to:

Gary Gentile Productions
P.O. Box 57137
Philadelphia, PA 19111

Additional copies may be purchased from the same address by sending a check or money order in the amount of $20 U.S. for each copy, postage paid.

Picture Credits

Unless otherwise specified, all photographs were taken by the author. The front cover view of the *California* is taken from a contemporary post card. The photograph in the upper right corner was taken in the small arms ammunition room of the *San Diego*. The rear cover photograph shows the author with a porthole he recovered from the *San Diego*.

The author wishes to thank Jon Hulburt for editorial advice, for valuable help and insight in making more precise the descriptive layout of the wreck, and for the use of his photographs; Colonel Richard L. Upchurch, U.S. Marine Corp (Retired), for access to the diary and memorabilia of his father, Able Seaman Otis Edward Upchurch; Paul Tzimoulis for his personal correspondence file; Ed Betts for taking the time for an interview; and Steve Bielenda for putting me in touch with Ben Manuella.

International Standard Book Number (ISBN) 0-9621453-1-9

First Edition

Printed in Hong Kong

CONTENTS

Introduction	7
Chapter 1: Construction	9
Chapter 2: Career	23
Chapter 3: U-156	47
Chapter 4: Sinking	55
Chapter 5: Salvage	67
Chapter 6: Preservation	83
Appendices	116
Books by the Author	120

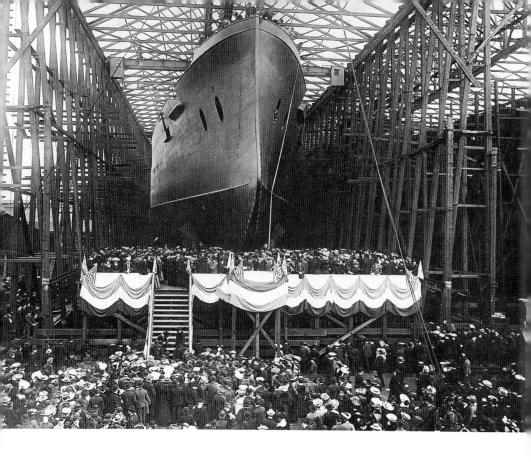

Above: The bow of the *California* on launching day. Note that the portholes have not yet been cut into the hull. Below: San Francisco Bay trials. Note that none of the broadside guns have been emplaced. (Both courtesy of the San Francisco Maritime National Historical Park.)

INTRODUCTION

A shipwreck is a time capsule: a fragment of history buried in the sea, a temporary repository of the remnants of a bygone age. The wood or steel hull is a transient abode that precariously extends the life of man's handiwork only slightly beyond the date of disaster and human suffering. This does not mean that the objects contained within are granted eternal life, for the sea is ever changing, ever destroying; it means only that oft sought relics have been granted a slight reprieve from obliteration.

To preserve a flag one does not hang it on a pole during a full gale: it is folded and packed away safely. One does not store precious china on an exposed mountain ridge where it is subject to rock falls, summer sun, and winter snow: it is kept in a glass case under controlled conditions. An artifact must be preserved *from* the elements of nature, not consigned to its capriciousness; the longer it is constrained to these wild forces the less likely it is to survive intact, to be found and appreciated by future generations.

On July 19, 1918 such a time capsule was born. When the *San Diego* sank within sight of American shores, she relinquished all claims to the title of warship and became instead a vast storehouse of marine equipment, naval stores, munitions, and the personal effects of over a thousand distressed sailors. She was a museum in the making; items then not worth the cost of salvage have slowly evolved into nautical antiques precious to the present-day generation.

The *San Diego* is a time machine: a shortcut to 1918, a slice of Americana. Once at war with the forces of man, she now fights a never ending battle against gradual but inevitable erosion by the chemical soup in which she is immersed.

To fully understand the *San Diego* we must travel back through the years not just to the moment of her birth, but to the time of her conception, for, although she lived in the twentieth century, she was a product of nineteenth century naval strategy....

Launching. (Courtesy of the San Francisco Maritime National Historical Park.)

CHAPTER 1
CONSTRUCTION

The concept of the armored cruiser is one that can best be understood in relation to contemporary tactics and technology. Battleships are heavily armed and armored, but slow and ponderous; cruisers are fast and maneuverable, but unprotected and outgunned by capital ships. The armored cruiser fills this gap by providing a compromise in armor and armament; it is made faster than the battleship by sacrificing the weight of big guns and heavy steel plate, and safer than the cruiser by offering a partial armor belt for the hull and decks: much like a vault inside a bank.

The Russians built the first armored cruisers in the 1870s. The British quickly followed suit, and in the 1880s so did the French. To a certain extent the designation of armored cruiser is a matter of opinion, based on design performance in relation to ships previously designated as battleships and cruisers. In that light, although the *Maine* was initially slated to become the U.S. Navy's first armored cruiser, while still under construction naval engineers decided she did not fit the category.

At seventeen knots the *Maine* was considered too slow, her ten-inch main guns too large, and her twelve-inch armor plate too thick. The Navy established the designation second-class armored battleship, and relegated ACR-1 to nonexistance. Thus, it fell to the *New York* to become the first U.S. ship to earn the title of armored cruiser.

The *New York* was authorized by Congress in 1888, and commissioned in 1893. She was 384 feet in length, displaced 8,150 tons, and carried four 8-inch guns as her main armament; during trials she maintained a speed of 21 knots. The *Brooklyn* was authorized in 1892, and commissioned in 1896. She was 402 feet in length, displaced 9,215 tons, and carried eight 8-inch guns; trial speed was 22 knots.

It was not until March 3, 1899 that Congress authorized a fleet of armored cruisers to be known as the *Pennsylvania* class. At first only three were authorized (*Pennsylvania, West Virginia,* and *California*), but a year later three more were added to the docket: *Colorado, Maryland,* and *South Dakota.* The overall length of 504 feet varied by as much as a foot, depending upon the shipyard in which the vessel was constructed. Displacement tonnage averaged 13,680 tons normal load, although this figure changed throughout each ship's career as alterations were made; full load displacement was just over 15,000 tons. Main armament consisted of four 8-inch guns in two turrets, bow and stern.

Strangely, these cruisers were some sixty feet longer than the U.S. battleships of the day (*Virginia* class), and carried more men. Their 22-knot speed made them three to four knots faster. Yet their full load displacement was only nine hundred tons less than that of a battleship.

The *Tennessee* class armored cruiser, authorized in 1902, and consisting of the *Tennessee, Washington, North Carolina,* and *Montana,* was the end of U.S. armored cruiser evolution. These ships were the same length as the *Pennsylvania* class, but broader abeam and of larger displacement, and carried four 10-inch guns as main armament.

Hindsight shows that the *Pennsylvania* class armored cruisers were obsolescent by the time of their construction. Naval doctrine intended these cruisers to serve as solitary units against commerce raiders and raider destroyers, and, in fleet maneuvers, to screen the sluggish battleships from fast enemy gunboats and torpedo boats, and to scout ahead of the main fleet.

During the 1904 naval engagements in the Russo-Japanese War, 6- and 8-inch guns proved useless. Although virtual clouds of these shells descended upon enemy ships having only long-range weapons, no serious damage resulted because these middle-weight projectiles did not have enough clout to penetrate the armor of the large-gunned warships; damage was superficial. Yet, a hit from a single 12-inch shell proved devastating to small-gunned and lightly-armored ships.

By 1905, when the *West Virginia* became the first of her class to undergo trials, the dreadnought era had arrived. Dreadnoughts were British battleships with all big guns, and initiated a strategy aimed at overwhelming warships having mixed-caliber guns: even if smaller guns could be brought close enough to bear, the damage they inflicted was insignificant. In this sense small guns were not only useless, but, because they took up space better occupied by larger guns, they were a liability.

To make matters worse, U.S. armored cruisers were powered by coal-burning reciprocating steam engines whose pistons and other moving parts could not take the wear and tear of continuous high-speed revolutions: they tended to fly apart under stress. The dreadnoughts steamed along effortlessly on clean, oil-fired turbines.

Yet, despite these drawbacks, the Big Ten armored cruisers served admirably in a variety of tasks, not the least of which was serving as a deterrent against foreign aggression by creating a military presence. Fitted with admiral's quarters they acted as flagships on distant shores: the silent vanguards of a fleet in waiting.

With the advent of the aeroplane, several armored cruisers were affixed with flight decks for launching light observation planes, or fitted with cranes for recovering seaplanes; these experimental exercises led to the development of the aircraft carrier. During the Great War the armored cruisers were used in the European theater for convoy duty and as troop transports.

The ship known today as the *San Diego* began her life as the *California*.

On January 10, 1901, the Union Iron Works, in San Francisco, was awarded the $3,800,000 contract to build the *California* according to Naval designs yet to be drawn. This price did not include "armor, ordnance, boats, portable furniture and lesser outfit usually furnished by the Government," which would cost an additional $1,000,000. The Union Iron Works was admirably suited to build the warship, as the cruiser *Olympia* had been built there between 1891 and 1895.

The keel was laid on May 7, 1902, from which date the company was required to complete the ship in the three years. However, six years passed before the *California* was finally accepted by the Navy. Part of the lagging pace was due to dockyard strikes, one in 1904 and another in 1905, for which the builder was not held accountable. A delay in the delivery of armor held up construction for a further eight months. Change orders and revisions added further postponement. Then came the prolonged strike of 1906, and the reorganization of labor forces after an agreement was reached. In addition, much of the material had to be transported by ship or rail from eastern manufacturers.

The crowning hindrance was the San Francisco earthquake of April 18, 1906, which devastated the city with tremors, collapsing buildings, and subsequent fires. Since the *California* was launched in 1904, she survived unscathed while nearby structures were shaken to the ground. In the massive clean-up and rebuilding program that followed, work on the *California* was largely set aside.

Originally, the *California*, as well as the *Pennsylvania* and *West Virginia*, was supposed to be sheathed in wood and copper as a deterrent against barnacles: those obnoxious marine shellfish whose attachment to ships' hulls created conical protrusions that increased hydrodynamic drag and diminished speed. Most ships intended for use in tropical waters were sheathed in order to reduce the amount of time spent in drydock for scraping.

Objections to sheathing were raised by the Board of Construction for a variety of reasons. The East India teak used as lath was not grown in the United States, and was not easy to obtain; attaching the wood to the armor belt was a difficult and laborious procedure with which shipbuilders had no experience. Later hull repairs would be more costly and complicated. The four hundred tons of additional weight meant a compromise elsewhere. Eventually, the Navy decided to save itself half a million dollars by not sheathing the three vessels.

The overall length of the *California* was 503′ 11″, her extreme beam was 69′ 10.5″, her normal draft was 24′, and her full load draft was 26′ 6″. One hunderd twenty-four athwartship frames, laid at four foot intervals except for the forward and after six frames, which were laid at two foot intervals, strengthened the bottom hull from bow to rudder pintle. Double-hull construction between frames 20 and 102 divided the ship into 279 watertight compartments. For water line protection, 22,400 pounds of corn pith was packed between the inner and outer shells; since corn pith swelled when wet, this acted as a self sealing agent in case of cracks in the hull.

Pneumatically operated doors that could be controlled electrically from the fore and aft conning towers could seal off the ship at a moment's notice. Each conning tower also had it own armored control station from which the ship could be conned.

Armor plate did not cover the entire ship, but only selected portions and then in varying degrees according to need. The plan view looks like a patchwork quilt of square, overlapping fish scales. Specially drawn Krupp steel six inches thick had the same armor-piercing protection as fifteen inches of common iron, and it was mostly with this type of steel that the *California* was armored.

As protection against torpedoes and low-trajectory shells, a seven-and-a-half-foot high armored belt girdled the ship from bow to stern, and extended up one foot above the deep-load water line. Along the engine and boiler rooms the steel was six inches thick, but forward and aft it tapered to three and a half inches. Amidships, five-inch armor extended upward to cover the 6-inch guns on the main deck. Four-inch transverse bulkheads at each end created a box effect. Along with the armored deck this formed a citadel much like an inverted shoebox without a lid. The 8-inch turrets and barbettes were also armored.

The hull was launched without machinery. Two years were required to install the two four-cylinder, tiple-expansion vertical engines, each in its own watertight compartment. Sample weights of parts were:

Low pressure pistons	3,489 pounds
Intermediate pressure pistons	3,009 pounds
High pressure pistons	1,933 pounds
Piston rods	2,411 pounds
Crossheads	2,339 pounds
Connecting rods	7,165 pounds

Steam was supplied by sixteen Babcock & Wilcox water-tube marine boilers in eight watertight compartments; a center-line bulkhead separated the port and starboard pairs of boilers, although each pair was connected to a common smoke funnel. A steam pressure of 250 p.s.i. could make 120 engine r.p.m. and develop 23,000 horsepower, enough to drive the ship at 22 knots. Sixty-nine-inch diameter fans mounted in the boiler rooms could blow air over the coals in order to accelerate the speed of burning, a system known as forced draft.

On her trials the *California's* boilers "produced more steam than the engines needed to make contract speed." Consumption was rated at 47,190 pounds of coal per hour, although this amount varied according to the type and grade of coal used. "The boilers supplied steam so easily that the work

Without machinery and armament the incompleted *California* rides high. (Both courtesy of the National Archives.)

of the firemen was light, and the air of the firerooms was very comfortable and not laden with dust. The smoke leaving the funnels was not heavy, even at its maximum." It was determined that 122.85 engine revolutions were needed to develop 22 knots. Maximum collective horsepower of both engines registered 28,942.9.

Twenty-eight coal bunkers with a total capacity of 2,000 tons allowed for an endurance of 5,000 miles at a speed of ten knots. Hoists lifted ash from the firerooms to the gun deck, and dumped it into overhead trolleys which then discharged the ash over the side. In port, ash lighters would take as much as six tons of ash at a load, and dump it in the open ocean. With todays environmental concerns, the EPA would look askance as such an operation.

Steam generated by the main boilers was siphoned off to operate dynamos and auxiliary equipment, such as fire and bilge pumps, ash-hoisting engines, ammunition hoists, and evaporating plants. The four evaporators and the distillers could make 20,000 gallons of fresh water daily. The ice machine not only made three tons of ice each day, but cooled drinking water in the scuttlebutt and the air in the magazines, as well as supplying low temperatures to the freezing tanks and cold-storage rooms. Steam heat from the engine exhaust was piped throughout the ship according to need. The windlass was powered by its own engine.

Seven generators supplied electricity at 125 volts in order to operate the motor-generators for gun elevating, a dynomotor for reserve battery storage cells, telephones (transformed down to 13 calling volts and 24 talking volts), shipboard lighting and auxiliary motors.

Lighting fixtures consisted of 319 steamtight lights, 294 bulkhead lights, 152 watertight portables, 126 battle lanterns, 95 ceiling fixtures, 72 desk lights, 69 bunker lights, 36 deck lanterns, 33 magazine lanterns, and peripherals such as masthead and side lights, signal lanterns, and parabolic reflector searchlights.

Motors operated such devices as electric fans (46), ammunition hoists (39), watertight doors (30), turret turning, rammers on the 8-inch guns, torpedo air compressors, boat cranes, deck winches, fresh-water pumps, dish washer, dough-mixing machine, laundry, and the engineer's workshop.

Wireless communication was then in its infancy. The *California* was equipped with a set whose range was less than 200 miles.

The two three-bladed propellers were made of manganese-bronze. Each was eighteen feet in diameter with a five foot hub, and weighed some 34,000 pounds. The pitch of the blades was adjustable. The two propeller shafts were each 48 feet long and 18.5 inches in diameter.

The main battery consisted of four 8-inch, 40-caliber guns, and fourteen 6-inch, 50-caliber guns. (Caliber, when stated along with barrel diameter, defines the ratio obtained by dividing the length of the barrel by the diameter of the bore. Thus, the barrel of an 8-inch, 40-caliber Naval gun is 320 inches long, or 26' 8", whereas the barrel of a 6-inch, 50-caliber gun

Cage lamps are often found in the wreck with their globes and bulbs intact. The ceiling fixture at lower right illuminated a corridor in the officers stateroom area.

is 300 inches long, or 25'.) These guns were fired by either electric or percussion primers.

For smaller armament the *California* carried eighteen 3-inch guns, twelve 3-pounders (so named because of the weight of the powder charge), two 1-pounder automatics, two 3-inch rapid-fire Hotchkiss field guns, two Gatling guns, and two Colt automatic rifles. All secondary battery guns were triggered by percussion primers only.

Ammunition was stowed in temperature controlled magazines and shell rooms. Every single day of every Navy ship's career, the magazines were inspected and the temperatures recorded. Spontaneous combustion igniting gun powder could destroy a ship, as it did to the USS *Maine* in Havana Harbor, in 1898.

Because no sparks could be allowed in the presence of so much explosive gun powder, the rooms were illuminated through cast bronze light boxes mounted through the bulkhead; electric lanterns could thus be isolated from the room, and serviced by a door which opened into the outer corridor. In the evolution of the cruiser design, the *Brooklyn* was the first to be fitted with two stage ammunition hoists that decreased the danger of flashbacks from the guns into the magazines.

For ammunition the *California* carried four types of shell: armor piercing, common, shrapnel, and blind (for target practice.) Quantities were calculated in rounds per gun: 125 for each 8-inch, 200 for each 6-inch, 250 for each 3-inch, 500 for each of the lesser guns.

The 3-inch and smaller guns fired fixed ammunition: a self contained unit consisting of a brass cartridge case and a projectile clamped together by a brass compression ring. The 6- and 8-inch guns used separate ammunition in two parts: a powder charge and a projectile. The powder charge is a silk bag filled with gun powder, and is stored in an airtight tank. The 6-inch guns used one powder charge per round, the 8-inch guns used two. The weights of the gun powder in all cases is carefully measured in order to ensure accuracy in firing.

The after 6-inch magazine (top) and shell room (below). Because the wreck lies upside down, the carved powder-bag tank-rests are now overhead.

Two stationary 18-inch torpedo tubes were fitted through the hull under the protective deck just ahead of the forward 8-inch magazine, for below-water discharge. Both torpedo tubes were located in the same watertight room, this room having no longitudinal bulkhead. Due to the narrowness of the hull, one tube was mounted slightly forward of the other because their lengths overlapped.

Bliss-Leavitt torpedoes that ran on turbine engines existed at the time, but the *California* was equipped with eight Whitehead torpedoes propelled by three or four cylinder reciprocating engines. Each torpedo was sixteen feet long, carried 140 pounds of wet gun cotton, and could run at twenty-eight knots with a maximum range of 4,000 yards.

Top: Torpedoes disassembled on the deck undergo routine maintenance. Below: The *California's* torpedo room, showing the tube and spare torpedoes. (Both courtesy of the Naval Historical Center).

Although most of the woodwork, both decking and furniture, was fireproofed, the use of so much combustible material on a warship came under severe attack by many ship constructors. Most of the gun deck and berth deck was covered with linoleum, except for wash rooms, toilet rooms, and passages, while the main deck and bridge were laid with teak. This may have looked good in the finest tradition of peacetime naval etiquet, but it was a decided disadvantage under fire, when a ship could become a seething cauldron from the hit of an enemy shell.

The *California* carried twenty-one boats, "composed of one 50-foot picket boat, two steam cutters of 36 and 33 feet, two 36-foot launches, five 30-foot cutters, one 30-foot admiral's barge, two 30-foot whaleboats, one 30-foot whaleboat captain's gig, two 20-foot dinghies, two dinghies of 16 and 14 feet, one 12-foot sidecleaning punt, and two Carley floats. The outfit was worked by two 15-and two 10-ton gooseneck cranes and eight sets of main deck davits."

Official trial runs were carried out on October 4, 1906, at which time the *California* ran up and down the Santa Barbara Channel fourteen times. "The ship vibrated very little, and no vibration was noticeable on the upper deck over the engine and fireroom spaces at any speed."

The only problem encountered was overheating of the crank-pin brasses at high speeds. Centrifugal force interfered with gravity-fed oil delivery resulting in loss of lubrication. This problem recurred during the October 11 four-hour speed trial, which had to be stopped after two and a half hours. The brasses were refitted before the October 22 speed trial, but the result was the same. After the failure of the October 26 speed trial "a centrifugal oiling arrangement was fitted to each main crank pin, which conveyed oil from a supplying receptacle to an annular grooved casting fastened to one web of each crank, concentric with the shaft. This casting was so fitted that centrifugal force impelled oil through a pipe from the casting into the hollow of the pin, and through the pin radially outward to the bearing surface." The November 12 trial proved the efficacy of this method of lubrication, and no more trouble was experienced.

At 10:26 a.m., on August 1, 1907, more than eight years after Congress authorized her construction, Armored Cruiser Number 6 was commissioned into the United States Navy. With her men mustered on the quarterdeck, Rear Admiral H.W. Lyon, Commander of the Mare Island Navy Yard, officiated the ceremonies. Captain Thomas Phelps, Jr., a seasoned veteran, proudly assumed command of the USS *California*.

Although she was not yet ready for sea duty, or the vicissitudes of war, on that day the *California* commenced her service for the United States Navy, to carry on a tradition inaugurated by rebels during the Revolutionary War. She, too, would know the meaning of war—but that was eleven years in the future.

Opposite: Construction phases. (Courtesy of the National Archives.)

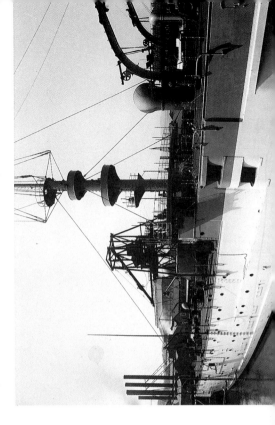
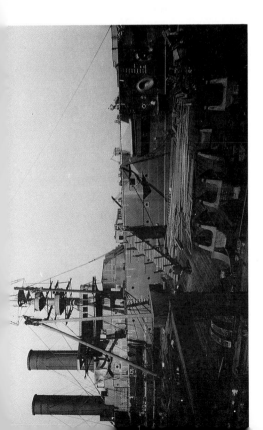
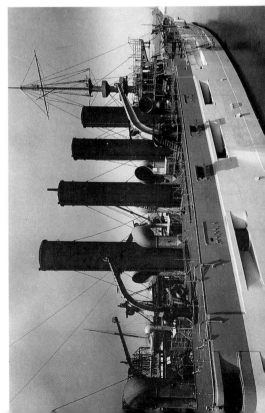

U. S. S. CALIFORNIA.

CHAPTER 2
CAREER

Big ships move sluggishly away from the dock. Tremendous power is required to impel such an immense bulk against the friction of the sea. Likewise, ships need time to gather momentum in forging a career.

Although officially registered as a vessel of the United States Navy, the *California* had a long way to go before she was prepared to meet the obligations for which she had been constructed. Below decks, engineers from Union Iron Works added the finishing touches to the machinery and power plants. Daily arrivals of men assigned to the newly commissioned duty station had to be quartered, and broken in to their new jobs. Artificers in all fields of shipboard maintenance familiarized themselves with the vast warship's most intimate workings.

In addition to the officers needed to handle the administration of the cruiser's tactical complexities, the *California* required the services of such military specialists as boatswains, yeomen, gunners, machinists, electricians, carpenters, shipfitters, shipwrights, oilers, plumbers, printers, buglers, bakers, bandmasters, blacksmiths, boilermakers, cooks, stewards, water tenders, painters, musicians, mess attendants, firemen, pharmacists, coppersmiths, and hospital apprentices: 850 men all told. Each night they were counted for bed check, and each morning they were mustered on deck for inspection.

Seamen of all eras are seamen, and American sailors are no different. As an example of the kind of difficulties encountered by Captain Phelps during his initial command of the *California,* he awarded weekly punishments at the mast for a wide variety of offenses such as shirking, neglect of duty, out of uniform, smoking out of hours, talking in ranks, profane language, leaving post without permission, shoes not shined at quarters, striking another person, late hammock, dirty accouterments, disrespect to non-commissioned officers, disobedience of orders, talking on post, gross carelessness (dropping a bucket overboard), dirty rifle, not falling in at quarters, wearing civilian clothing ashore, making unnecessary noise after pipe down, spitting, gambling, being dirty, and concealing venereal disease.

The log was appended with a typed list of all those receiving extra duty, fines, reductions in rank, or solitary confinement in the brig wearing double irons and fed nothing but bread and water. As they say, those were the days.

Opposite: The *California* on a contemporary postcard.

On August 10, Able Seaman H. Schnell was delivered on board by civil authorities 194 hours overdue from liberty. He was AWOL for eight of the nine days of the *California's* time in commission. It was with this kind of behavior that Captain Phelps was forced to deal, as was every other commanding officer in any navy in the world. Boys will be boys, and sailors will always be sailors.

The only item of interest during the *California's* first month in service was a visit by the Secretary of the Navy. He inspected the ship on August 29 on a perfunctory basis.

On October 12, Captain Phelps relinquished command of the *California* in order to assume directorship of the Mare Island Navy Yard, at Vallejo, California. His executive officer Lieutenant Commander Newton McCully took temporary command. McCully's brief stint lasted only until after the acceptance trials. November 13 found the *California* leaving San Francisco for Puget Sound, where she was put into dry dock at Bremerton, Washington.

Captain Vicendon Cottman assumed command on November 18, and McCully went back to his position as executive officer. While Cottman acquainted himself with his men and his ship, the *California* was scraped free of barnacles and had her hull painted. Massive leaks were discovered in the armor. Most of the zincs were replaced in the never ending task of cathodic protection. Repairs were accomplished during the week in dry dock.

Cottman brought the *California* back to her home state; December 17 dawned with the ship anchored off Vallejo. However, the crew was not to enjoy Christmas in home port, for on the eighteenth the cruiser took a jaunt along the West Coast as far south as Magdalena Bay, halfway down the Mexican portion of the Baja Peninsula. She did not return to San Diego until January 5, 1908.

After that, she participated in the unveiling of the *Bennington* memorial, which commemorated the ship on which a boiler explosion killed sixty men and wounded forty. It was a strange premonition of events to come.

After her shakedown run south of the border, the *California* required some engine overhauling. This was accomplished at Union Iron Works. On January 18, she left for her final trial trip, during which she fired her guns for the first time. Teething problems arising from the trials were only what was expected. By February 20 she was ready for another trip to Magdalena Bay, this time for target practice. Then she returned to San Francisco.

When May 14 arrived so did the *California's* final acceptance, although it was not until August 15, 1909 that Union Iron Works received final payment in a long list of vouchers that began in May of 1902. If there was any celebration aboard the cruiser, it was not entered into the log. In fact, acceptance is a condition that is visible only on paper, not on deck. On the seventeenth she participated in the naval review for the Secretary of

Above: A view of the bridge. Below: The view from the bridge. (The original glass plates are cracked, and have been taped. Courtesy of the National Archives.)

the Navy. Perhaps more notable to the men was the discovery by the medical officer that 150 pounds of macaroni was bad and had to be thrown overboard.

The *California* was assigned to the Second Division of the Pacific Fleet. During the summer months of 1908 she engaged in squadron activities with sister ships *South Dakota, Washington,* and *Tennessee,* steaming along the Pacific coast and stopping in port cities for exhibition purposes. The new, sleek four-stackers, in their buff and white color schemes, were the pride of the fleet. Judging by the increased numbers of seamen receiving punishment, a good time was had by all.

The fun and festivities ended on August 24, when both divisions of the First Squadron left for the long ocean voyage to Samoa. Together with three armored cruisers of the First Division (*Pennsylvania, Maryland,* and *West Virginia; Colorado* was having her 8-inch guns replaced), seven destroyers (each of which had to be towed by a cruiser because of the destroyers' limited coal-carrying capacity), and the hopsital ship *Solace,* they presented a formidable military force.

California towed the *Truxtun,* which had accompanied Theodore Roosevelt's Great White Fleet from Hampton Roads, Virginia during the first leg of its around-the-world journey. (In May, when the Great White Fleet had steamed across the Pacific, the *Truxtun* had been left on the West Coast. The Fleet consisted mostly of battleships; armored cruisers were not part of the circumnavigation.)

In the log for September 16, while in the middle of the Pacific Ocean the Officer of the Deck noted matter-of-factly, "Neptune and party came aboard at 8:30 and were received with customary ceremonies after which land lubbers were initiated into ceremonies of crossing the line."

The crossing of the equator resulted in the usual shipboard shenanigans associated with introducing "pollywogs" into Neptune's realm. All those

Gun crew at practice. (Courtesy of the Naval Historical Center.)

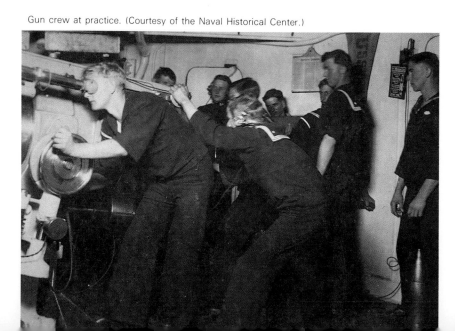

who had never before crossed the zero latitude into the southern hemisphere had their heads shaved, and were put through a series of initation rites which were as farcical and as preposterous as the imagination of the "shellbacks" who had already undergone such inanities upon their own first crossing, and who were more than willing to see others suffer the same fate. These incivilities might include body painting, mud baths, donning strange dress, and other assorted demeaning, boyscoutish antics. Officers were not immune to treatment, and had to endure the same ill adventures as the enlisted men.

The Fleet reached Pago Pago, Samoa, on September 20. For most of the men, however, the return passage through Hawaii was undoubtedly the highlight of the trip; the Sandwich Islands offered more opportunity for entertainment (and debauchery) than the desolate coastal communities of Samoa. The *California* pulled past Diamond Head on October 16, and left Hawaii six days later.

November 2 found the *California* and her squadron back in Magdalena Bay. They spent a month on target practice. Then, due to political unrest in neighboring Central and South American countries, they left almost immediately for another show of force.

On December 30, Able Seaman J. Bok "fell overboard from the starboard side of the forecastle while placing a clean cover on his mattress after hammock inspection." Life buoys were tossed after him, but floated by and were sucked under the cruiser's wake. The *California* pulled out of formation, lowered lifeboats, but by that time Bok had disappeared. Afterward, as the men were securing the front lifeboat, the shank of the swivel parted and two men were thrown overboard. Since the ship was still at rest, they were saved when two other men jumped to their rescue. All four were picked up by the starboard lifeboat.

Life aboard ship was always hazardous, as one would expect when

A similar gun on the USS *Olympia*, in Philadelphia.

fragile human flesh is constantly thrust against unyielding steel. Fatalities were rare, but injuries abounded. A ship is a great metal enclosure for all types of machinery. Broken bones were fairly common place, and amputations of digits nearly so.

January 3, 1909, found the *California* at Talcahuano, Chile. She later moved to Panama. In early March she anchored in Amapala, Honduras, and by the twentieth she resumed target practice in Magdalena bay. In port, the men played baseball and went swimming.

The *California* spent the spring patrolling the coast and "sharpening her readiness through training exercises and drills." She was in Tacoma, Washington the week of May 24, when two of His Imperial Japanese Majesty's Ships, *Soya* and *Aso,* arrived to celebrate the fourth anniversary of the Japanese naval victory over the Russians. Captain Cottman made an official call on the Japanese admiral during a reception aboard the *Aso*. American and Japanese ships fired salutes, each flying the ensign of the other nation.

On June 10 the *California* went back to Mare Island, where she spent two months being overhauled. Changes were made constantly; in 1908, automatic shutters were fitted in the ammunition tubes, the purpose being to lessen the danger of flashbacks from the turret's gun room down the tube into the magazine. The military foremast was removed and replaced with a lattice cage mast and fire control top. Spotters now directed ranges and elevations from new heights, and could more accurately aim their guns.

Dry-docking revealed two layers of barnacles a half inch thick: a living outer layer growing over a dead inner layer. Brown grass an inch long festooned the hull and the propellers. Most of the antifouling was gone. Scrapers toned their muscles during a week of hard work.

Below: In dry dock, men clean the *Californie's* propeller, (Courtesy of Colonel Richard L. Upchurch, U.S. Marine Corp (Retired). Opposite: Coaling was a dirty job, and so was the clean-up afterward. When the work was done the men had time for shark fishing. (All three courtesy of the Naval Historical Center.)

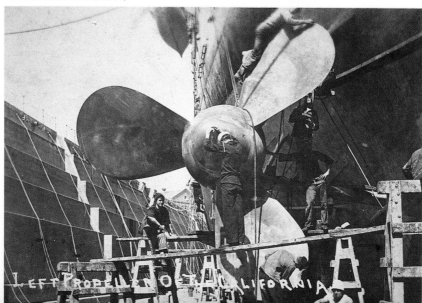

Captain Henry Mayo took command of the *California* on September 3, 1909, in time to take her on another Pacific jaunt, touching such places as the Admiralty Islands, Manila, Yokohama (Japan), and Woosung (China), before swinging back through Honolulu. By January, 1910 the cruiser was stateside. The pilgrimage to Mare Island for repairs became an annual event. Fleet exercises, gunnery practice, coaling, painting, and drill kept the men busy.

The main diversion occurred from mid-August to mid-October, 1910. After a stopover in Chimbote, Peru, the *California* visited Valparaiso in order to join in Chile's centennial celebrations. In times of peace a nation's warships often took the role of strengthening diplomatic relations.

January 17, 1911 was a banner day for the *California,* for Rear Admiral Chauncey Thomas hoisted his flag on her mast and designated her the flagship of the Pacific Fleet. Captain C.H. Harlow relieved Captain Mayo of command. Most of the year passed quietly for ACR-6, anchored at Mare Island. During the Independance Day celebrations her boat crews demonstrated their skill by winning the six-oar and twelve-oar cutter races. A championship trophy cup was presented to the winners by the city of Vallejo.

Six-oar racing trophy, recovered and photographed by John Lachenmeyer.

When the *California* entered Pearl Harbor she broke a ribbon stretched across the harbor mouth. Notice the canopy on the after deck; it provided shade from the tropical sun. Also, a wire cage now encircles the forward mast.

On November 22, the *California* led the squadron on a cruise across the Pacific Ocean. First stop was Hawaii. Admiral Thomas wrote, "USS *California* yesterday entered the new channel dredged across the reef and passed with safety into the inner harbor, anchoring off the entrance to the dry dock under process of construction.... The object of this visit was to test out practically the work recently completed, and in this manner to demonstrate the navigability of the channel." He added that the passage "was without significance in so far as appertained in any way to an official opening of the naval station or its harbor ... that the sole object was to ascertain that the channel was navigable." The harbor he referred to was the now-famous Pearl Harbor; the *California* thus became the first rated man-of-war to enter the new establishment.

When Captain Harlow reported for a week of Board duty in connection with the progress of work at Pearl Harbor, Lieutenant Commander R.S. Douglas took temporary command of the ship.

In 1912, after several months of leisure, the *California* continued her westward voyage to Asia, putting in at ports on the islands of Guam and in the Philippines. Captain Harlow retired from the Navy on April 14, leaving command to Lieutenant Commander Willis McDowell for two days, until the arrival of Captain C.M. Fahr. Two months later, the *California* left for Woosung, where Captain A.S. Halstead took permanent command of the vessel.

On July 8 the *California* was docked in Tsingtau, China, when the German cruiser *Emden* steamed out of port. Later, during the Great War, the *Emden* became a famous merchant raider in the South Pacific, sowing a path of destruction that forced the entire British Pacific fleet into searching for her. After many months, she was finally tracked down and sunk by the HMS *Sydney*.

The *California* returned to the contiguous states in August, docking in San Francisco, but stayed only briefly before being dispatched to Corinto, Nicaragua, carrying a Marine detachment for immediate deployment. This was no diplomatic mission, but a real war patrol, under the overall command of Rear Admiral W.H.H. Southerland, who flew his flag on the *California* during the operation.

Political instability wracked the small Central American country, endangering American lives and property. When Colonel Joseph Pendleton took his Marine regiment ashore, the *California's* landing force went with him. Twenty-seven hundred men laid seige to the capital city of Managua. During fierce fighting, rebel forces under General Mena were overthrown, and two gunboats were captured. The *California* did not actually enter the fracas, but offered logistical support.

After quelling the revolution the *California's* next duty station was Mexico. During 1913 she kept a close eye on the politically disturbed country. Newton McCully, now a captain, took command on August 21.

The *California* spent practically all of 1914 anchored in the mainland port of Mazatlan, occasionally steaming off to other troubled Mexican cities. She did not become involved militarily, but acted merely as an observation vessel. As Federal forces battled rebel gunboats, and were chased out of the harbor by the insurrectionists, the *California* floated serenely nearby.

Additional insights are evident in a private diary kept by Able Seaman Otis Edward Upchurch, who saw the Mexican civil war from the deck of the *California*. He wrote with repressed frustration that "she expended a lot of powder saluting Mexican gunboats."

March was a highlight in the *California's* career. During a twenty-four hour full-power test run she broke all records ever made by any battleship or armored cruiser.

Back at anchor off Mazatlan on August 8, Upchurch noted grimly in his diary, "Six dead bodies came floating by, also two or three heads." For the Mexicans this was full scale war. Still, the U.S. was merely an observer, and did not involve itself with internal political machinations. The *California* became a harbor for fleeing refugees, at times quartering as many as a hundred.

Deck log, August 5: "Captain N.A. McCully was detached from command of this ship and left ship in obedience to telegraphic orders from the Navy Department dated August 4, to proceed to Washington D.C. and report for temporary duty in the Office of Naval Intelligence preparatory to proceeding to Russia as attache."

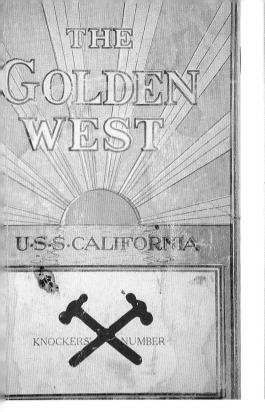

THE GOLDEN WEST

PRINTED AND PUBLISHED MONTHLY ON THE U. S. S.
CALIFORNIA, FOR AND BY THE CREW.

VOL. 1.   APRIL, 1914.   NO. 4.

CRUISING

We did considerable steaming last month visiting, in addition to Mazatlan, the ports of Guaymas and Topolobampo, Mexican ports in the Gulf of California.

Shortly before eight o'clock on the morning of Thursday, March 5th, we got underway at Mazatlan bound for Guaymas on engineering trial runs. About one-half hour later, having sighted our Collier Nanshan standing down from San Francisco, we heaved to to receive Paymaster Schumann and First Lieutenant Vulte and seven sacks of mail, the Paymaster and Mr. Vulte having been granted brief leaves of absence just prior to our departure from San Diego a few days before and were rejoining the ship. Had we waited until the Collier got within three miles of shore we could not have received the officers and mail from her until after the port Doctor had made his visit and raised her quarantine. Intercepting her beyond the three mile limit, her status being that of a ship on the high sea, saved us considerable time.

After our whaleboat with officers and mail had been hoisted in we continued on our course, steaming under sixteen boilers, forced draft, making full power run. Throughout the day and night and during the forenoon of the following day the members of the Engineer's force were bringing to bear every bit of skill and energy they could muster, the water tenders, firemen and coal passers making the steam and the machinist's mates and oilers using it to the best advantage. Not only must the engines be given sufficient steam to keep them turning over the maximum number of revolutions but all parts must be carefully watched to guard against their working loose or becoming heated. Throughout the full twenty-four hour test there wasn't a single mishap, and the good ship broke all former records for Battleships and Armored Cruisers of our Navy, thanks to the most excellent skill of the "Black Gang", from the Chief Engineer down.

During the early morning of Friday, March 6th, Clarence E. Coyner, ordinary seaman, a native of Bolekow, Missouri, was reported missing, it being brought out later by Boards of Inquest and Investigation that he leaped over the ship's side with suicidal intent supposedly while in a mentally deranged condition. The night being dark and with the ship at high speed there was little chance of his act being detected at the time.

We arrived off Guaymas at about six o'clock Friday evening, where were the Raleigh, the Japanese cruiser Idzuma and our Collier Justin. The following morning was devoted to cleaning the ship thoroughly as the cinders and coal dirt in general had been very much in evidence throughout the speed trial just completed, forced draft through the four large funnels causing a perpetual falling to the decks of cinders ranging in size from the Pullman-sleeper variety to those of the *heft* of *Irish confetti*, (or so it would seem should one happen to strike you in the region of the optic.)

Guaymas is another of those quiet little Mexican sea-port towns (15000), surrounded on three sides by barren mountains some 1500 feet in height, affording excellent protection from invading forces in time of revolution. The bay is land-locked, dotted with small islands and shallow, permitting entrance of light draft vessels only.

We shall ever remember Guaymas by the beautiful mirages seen in its vicinity almost daily. Each morning, the water being perfectly smooth, reflections of nearby islands were visible as clear cut as though cast in a mirror, producing an optical effect that beggars description. Due to the reflection of light through differently heated strata of air the objects assumed ever-changing proportions.

On the 14th we left Guaymas for Mazatlan, stopping for a few hours off Topolobampo, a very small place unworthy of note.

You've had your beans and coffee cake,
   And your mug is minus pain—
It's a dead sure sign the knockers
   Will all ship over again.—*"Bill" Harris.*

A SAILOR'S PRAYER
(With apologies)

Now I lay me down to sleep,
I pray thee, Lord, my soul to keep;
   Grant no other sailor'll take
   My shoes or socks before I wake.
O, Lord, guard me in my slumber
And keep my hammock on its number.
   May no clews nor lashings break,
   And let me down before I wake;
Keep me safely in thy sight,
And grant no fire-drill in the night;
   And in the morning let me wake
   Breathing scents of sirloin steak.
God protect me in my dreams,
And make this better than it seems.
   Grant the time may swiftly fly,
   When myself shall rest on high
In a snowy feather bed,
Where I long to rest my head;
   Far away from all these scenes,
   From the smell of half-baked beans.
Take me back into the land
Where they don't scrub down with sand;
   Where no demon typhoon blows,
   And where the women wash the clothes.
God thou knowest all my woes,
Protect me in my dying throes.
   Take me back—I'll promise then,
   Never to leave home again.

* * *

Our Father who art in Washington!
   Please forget about the past—
   The things that happened at the mast.
I'll promise to obey each rule
And never miss a day at school!
   Do not my request refuse—
   Let me stay another cruise.

THANKSGIVING-DAY DINNER
1915

FORMER U.S.S. CALIFORNIA FLAGSHIP
U. S. S. SAN DIEGO
PACIFIC FLEET. I WAS ON BOARD. C.E.V.
FRANCISCO . . . . CALIFO

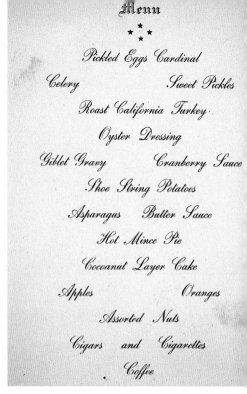

Menu

Pickled Eggs Cardinal

Celery   Sweet Pickles

Roast California Turkey

Oyster Dressing

Giblet Gravy   Cranberry Sauce

Shoe String Potatoes

Asparagus   Butter Sauce

Hot Mince Pie

Cocoanut Layer Cake

Apples   Oranges

Assorted Nuts

Cigars and Cigarettes

Coffee

MUSIC PROGRAM

MARCH
My Croony Melody

TWO-STEP
I Love the Ladies

SONG-MARCH
It's a Long way to Tipperary

TWO-STEP
When it's Night Time Down in Burgundy

TWO-STEP
Some Baby

RAG
Ragamuffin

THANKSGIVING D
U. S. S. SAN DIEGO
1914

Able Seaman Otis Edward Upchurch captioned this picture as "Mexicans selling their wares." (Courtesy of Colonel Richard L. Upchurch, U.S. Marine Corp (Retired).

Lieutenant Commander J.H. Holden, the Executive Officer, wore the captain's hat until August 13, when Commander Ashley Robertson left the *Denver* to take command of the *California*.

On September 1, 1914 the *California* was relieved of the name she carried since her inception. Because of a new Naval policy of naming battleships after states, the recently authorized BB-44 was christened *California*. ACR-6 became the USS *San Diego,* after that port city practically on the border between the U.S. and Mexico.

Not all of a sailor's life was drudgery. Witness the simplicity of Upchurch's entry for December 20, after receiving pay. "Beach party all day." Pichilinque Bay was probably never the same afterward.

The *San Diego* was a happy ship, and offered her crew much in the way of entertainment. The men played cards and acey-deucy (a variation of backgammon), sang in barbershop quartets, or joined in smokers. (Smokers were not the stag films or porno flicks of modern day usage, but scoial gatherings where the men actually smoked while boxing, playing musical instruments, or reciting poetry.) The off-duty amusement most often mentioned in Upchurch's diary was watching silent movies on the quarterdeck.

In order to maintain her warlike readiness and to develop battle efficiency, the *San Diego's* gun crews often partook in target practice. Sometimes they used target-practice shells of cast steel or iron, filled with sand instead of a bursting charge, and fired them at towed wooden targets.

The spotting rifle in the upper left corner did not look like much when I first saw it, but because most of it was made of brass I was able to restore it nicely. The barrel brush was in the same compartment.

But usually, as a way of saving shells and powder, and to avoid having to clean the big barrels after each session, subcaliber guns were fired in place of the main battery.

The subcaliber gun, or spotting rifle, mounts on the barrel of a larger gun, and is fired by a lanyard. During gunnery practice the crew went through the timed loading drill, but instead of firing the main gun the gunnery officer pulled the lanyard for the spotting rifle. Either blind shells (blanks) were used, or a bullet catcher was secured in the line of fire to prevent accidental injury.

Tragedy struck the *San Diego* on January 21, 1915 while steaming across the Gulf of California, from La Paz to Guaymas, to receive coal. Captain Robertson took the conn during a four-hour full-speed and endurance run using all sixteen boilers under forced draft. When they hauled in the patent log they obtained a reading of 21.7 knots, proof that the cruiser's engines could still adhere to their designed specifications.

However, only minutes before the trial time was over, boiler number 2 blew out a tube. Ensign R.W. Cary, Jr., testifying at the Board of Investigation, stated "I was standing just inside the door in number 1 fireroom ... was facing directly toward number 2 boiler when I heard the sound of an explosion, the fire box doors all flew open and flame, coals and smoke came out of the doors. I stepped over to the lever of the power door and stood there waiting for men to come out of that fireroom." He sealed the door after the last man came through.

Moments later "the explosion of number 7 boiler occurred, perhaps with a little more violence than number 2, this brought a cloud of smoke into number 1 fireroom and a number of men escaped into number 1 from number 3 fireroom, then the door connecting number 1 and 3 firerooms was closed."

Charles Smith, a water tender, who was on top of number 8 boiler, stated, "I was oiling my blower when the explosion occurred and the gas blinded and choked me, and I passed over the gratings and escaped through the drum room, the only escape I had. When I got to the foot of the ladder leading to the drum room, the second explosion occurred. I came out through the drum room and don't remember anything after I came out through the drum room."

Albert Priddy, water tender, stated, "I was standing between the feed pump and the throttle of the blower on No. 7 boiler, and when the explosion went off, it blew me out into No. 5 fire room, the concussion was so great."

Water tender Walter Redmond stated, "I was going toward the after boiler and was nearly there when the explosion occurred. It looked like a coke fire being pulled and a salt water hose being turned on it. It was one big blur of smoke and steam. The explosion knocked Elliott against the front of the after boiler."

Frank Bush, chief water tender: "I was in charge of the fire room repair party, and I had just stepped into this fire room from No. 5. I noticed a flash like that from a gun. Everything became dark and I felt as if something had raised me up, and the next thing I knew, I was lying against No. 5 feed pump in No. 5 fire room. I gathered myself up and went directly back into No. 3 fire room. I saw no one there and the watertight doors were closed on me and I had to go up the ladder to get out."

Boilers No. 4, 6, and 8 blew out tubes as well. Twenty-three men were seriously scalded by expolding steam. Except for boiler No. 15, which was supplying auxiliary power, the fires under the rest of the boilers were damped. The *San Diego* went adrift. Within hours the burned men began succumbing to their injuries, and by the end of the day four firemen were dead. The next day a water tender passed away.

While the *San Diego* took on coal from the USS *Saturn*, a Board of Investigation was convened in order to determine the cause and extent of the damage, and whether the ruptured tubes could be repaired or replaced by the crew. Three days of testimony, inspection, and deliberation followed. The Board found:

"No's 1 and 3 main feed pumps lost their suction due to an obstruction

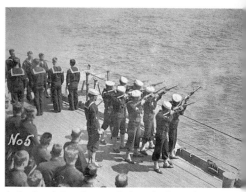

Burial at sea on the *San Diego*: services are read, the band dons hats and plays the funeral dirge, the coffin is tipped over the side, and a gun squad fires a final salute. (Courtesy of the Naval Historical Center.)

in the suction pipe. The top of the strainer over the only suction outlet from starboard feed tank came off and was drawn down inside the strainer, completely obstructing the opening to suction pipe."

Half of the condensed water was thus lost, while no fresh water could be pumped in. The boilers quickly overheated and exploded. Even as this determination was being made, two more men died. On the twenty-seventh there was another death, and on the thirtieth, another. Fatalities totaled nine.

Although the two least seriously damaged boilers could be repaired in fourteen days working time, the other six "are beyond the capacity of the ship's force to repair, as, in the opinion of the board, all headers that were overheated should be annealed, provided they are not blistered or ruptured."

The *San Diego* was laid up at the Mare Island Navy Yard until mid summer, and placed in reduced commission. On March 2, Robertson and the flag moved to the *Colorado*. For one hour Lieutenant C.C. Soule had temporary command, until the arrival of Lieutenant Commander B.G. Barthalow. Then, from March 23 to June 20, Commander C.D. Stearus was the nominal captain of the *California*. Stearus got to move the ship only once, to the coaling station at California City. Then Barthalow took back his command.

Meanwhile, local newspaper headlines for February 14, 1915 touted "Trophy for Gunnery Prowess is Given to *San Diego's* Crew." Secretary of the Navy Josephus Daniels offered his praise: "The Spokane Trophy, which is annually awarded the vessel making the highest merit with turret guns at elementary practice, is therefore awarded to the U.S.S. *San Diego* for the year 1914–1915."

In addition to the encomium, a monetary prize was awarded to the meritorious crews at the rate of $20 per man.

Upchurch's diary offers another insight into the dangers of being a sailor: "Carter was killed while coaling, 4 bags fell on him, another was injured."

Once back in full commission, the flag of the Pacific Fleet was returned to the *San Diego*. During the summer she lay at anchor off the Exposition Grounds at San Francisco, and joined in the festivities. All was not fun and games, however, according to Upchurch's diary entry for July 14: "Fire and collision drill for morning. Fourth Div. jumped over the side with life preservers on just after collision drill." It was this kind of preparation which was to save many men's lives three years later.

Robertson returned to the *San Diego* on September 15 for his second stint as her commanding officer; he was now a captain.

Left: The Spokane Trophy. (Courtesy of Colonel Richard L. Upchurch, U.S. Marine Corp (Retired). Right: The *San Diego's* canteen, where sailors could purchase food and drink not supplied by the Navy. (Courtesy of the Naval Historical Center.)

On October 13, the *San Diego* assisted the Triangle Motion Picture Corporation in filming "The Power of Solom Bay." The movie company took stock footage on the deck of the cruiser, and of the men at their gun posts, then, in order to recreate a battle at sea, built a seventy-five foot wooden replica of a battleship.

Upchurch described the action in his diary. "The movie men came aboard and took various scenes in the forenoon. The starboard battery fired at the movie target about 3:00 PM. Six shots were fired per gun. 2 pounds of black powder per bag were fired first to make more smoke. The ship failed to sink tho all mast and stacks were shot away. Several shots were put through the hull and she finally caught a fire. The range was about 2000 yards." A ship carved out of a block of wood, the director discovered to his dismay, is difficult to sink.

Both top and bottom courtesy of Colonel Richard L. Upchurch, U.S. Marine Corp (Retired).

Christmas in Guaymas. Notice the Christmas tree on the bow. (Courtesy of the Naval Historical Center.)

November 6 found the *San Diego* off the west coast when her radio operator intercepted a call for help from the wrecked steamer *Ft. Bragg*. *San Diego* raced to the rescue, and succeeded in taking off forty-eight passengers.

Mexico was again in uprising. In November, several ships of the Pacific Fleet, flagship *San Diego* especially, carried marines to help quell the disturbances. However, the marines were not landed; the *San Diego* maintained her military presence until February, 1916 when she took the marines home. She spent the rest of the year cruising the west coast. Command shifted to Commander George Bradshaw on June 8.

On February 12, 1917, by order of the Bureau of Navigation, the *San Diego* was placed in reserve status under the command of Commander Claude Price. She docked at Mare Island for overhaul, but this time she did not leave when repairs were completed. However, she did not stay in that reduced capacity for long. With the U.S. entry into World War One, she was recommissioned as flagship of the Pacific Fleet as of April 7. The muster rolls swelled as the men began coming back to their ship.

During May, twelve of her 3-inch guns were dismounted and turned in to the Navy Yard; this left only six. Three thousand rounds of extra 3-inch ammunition were also off-loaded. In its place came 600 rounds of 8-inch ammunition, 1,300 rounds of 6-inch ammunition, and 9,216 candy bars. The armament list also documented 350 .30 caliber Springfield rifles (1903) and 100 .45 caliber Colt pistols (1911).

Since there was no war on the Pacific side of the continent, July 19 found the ship underway to an undisclosed port listed in the log as X. Sealed orders took her to Panama, which she had visited many times before. This time, however, she entered the canal. On July 29, the ship that had never been out of the Pacific Ocean, tasted the salt of Atlantic waters. She ran darkened at night, with storm covers dogged, and arrived at Hampton Roads on August 4. A week later she arrived at the Brooklyn Navy Yard and anchored in the North River.

On August 19, Captain Harley Christy assumed command.

The *San Diego's* new assignment required her to escort convoys carrying troops and supplies to the beleaguered European front. Operating out of Halifax, New York, and Hampton Roads, she braved the stormy, U-boat infested North Atlantic in pursuit of the war.

During the course of convoy duty her armament was modified. First, depth charge racks were added to her stern; later, she was armed with two 3-inch antiaircraft guns and two Lewis machine guns; her Whitehead torpedoes were replaced with Bliss-Leavitts.

In keeping with a policy of wartime secrecy, observations usually made in the deck log were omitted. No longer was the destination written in, or the compass headings or engine revolutions recorded, or each watch's sextant sightings noted for the following watch. The men did not know where they were going until they got there. In fact, most of the time the *San Diego* returned to her port of origin without touching another port; yet she came back with a convoy different from the one with which she had started.

A mid-ocean rendevous was the trick. Convoys from Europe and America started out at approximately the same time, met at a prescribed location in the broad reaches of the Atlantic, and exchanged escort vessels. In this way the *San Diego* would ride shotgun on a fully loaded convoy halfway, then turn around and come back with a returning convoy in ballast. If her deck log fell into enemy hands, it would reveal nothing about her movements other than the fact that she had moved.

In November, the *San Diego* made her only complete crossing of the Atlantic Ocean, when she accompanied a convoy to La Croisic, France.

The *San Diego* continued escorting transatlantic convoys throughout the first six months of the year. On June 12, along with the USS *Plattsburg* and the HMS *Edinburgh Castle*, she left New York with the fast Mercantile Convoy HX-37. When the ten merchant ships arrived safely at Liverpool, England, the *San Diego* was already on her way back to Portsmouth, New Hampshire, which she reached on the twenty-ninth.

She entered dry dock the next day. The three central 6-inch guns on each side of the gun deck were off-loaded. The two pairs of forward gun ports were sealed with oak planking bolted through top and bottom with

Top: The *San Diego* buries her bow during a North Atlantic patrol. Middle: The *San Diego* is heavily iced over after a cold Atlantic crossing. Bottom: The *San Diego's* crew wear dress whites in anticipation of liberty. (Top and middle, courtesy of the Naval Historical Center. Bottom, courtesy of Colonel Richard L. Upchurch, U.S. Marine Corp (Retired).

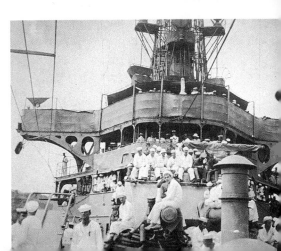

angle iron, then caulked with four threads of oakum, giving the hull a flush finish. The gun shutters were permanently closed over the other 6-inch gun ports; in order to make them watertight the holes through which the barrels had protruded were covered with iron plate and gaskets, and secured with strongbacks. The four 3-inch gun ports were closed with their own shutters, and the holes similarly sealed.

A freight carload of paymaster's stores came aboard, and was stored forward. The *San Diego* came out on July 16. Because she had orders to sail with a transatlantic convoy on July 22, she took on 2,938 tons of coal: more than her bunkers could accomodate. Extra coal bags were packed in corners, unused rooms, and on deck.

She headed south for New York on July 18, in the process very likely intersecting the path of a German U-boat that was working itself north. Indeed, due to the known presence of U-boats off the coast, Captain Christy issued orders that the *San Diego* steam a zigzag course in accordance with the Mercantile Convoy Instructions, pattern #10. She rounded Nantucket Shoals late that night, and turned west.

July 19 dawned warm and slightly hazy. As the *San Diego* steamed along the south shore of Long Island, she continually dropped her sounding lead as a way of determining her position by following the trend of the coast line; the navigator was unable to take a sun sighting.

The *San Diego* was in a complete state of battle readiness. According to Christy, "no compartments or storerooms below the water line not in use were allowed to be opened except by express orders of the Executive Officer and then only one at a time with men standing by to close them instantly in case of necessity. The automatic system for closing watertight doors was operated daily and defects were corrected immediately after their discovery. A thorough canvass of the ship was made for the purpose of discovering all holes left in bulkheads and all such holes were closed. As a part of the policy of maintaining our water-tightness an ample supply of broadside ammunition was kept on deck so that ammunition magazines would not have to be opened except in case the ship was required to engage with her turret guns."

Christy also stated that "some of the ventilating ducts in the storerooms and magazines were fitted with float valves which worked automatically, most of the voice tubes have mouthpieces fitted with a flap which can be clamped shut, but do not close automatically from water pressure."

Enough life preservers were stowed on the Boat Deck to equip every man aboard; canteens filled with fresh water were stashed nearby.

A continuous radio watch was kept on the receiving set, in case any vessels in the vicinity reported the presence of a U-boat. However, no watch was maintained on the submarine signal apparatus because Christy "did not consider it efficient on account of the water and other noises about the ship." Seventeen lookouts, each assigned to observe specific arcs, were positioned around the ship. Fire control parties and all the gun crews stood by their stations.

Only hours prior to her expected arrival in New York, as the *San Diego* steamed along quietly but fully alert, the warship was alone in the vast, placid sea. Her speed was fifteen knots. Captain Christy stood on the top of the wheel house, eight feet above the deck: a position from which he could see the bridge watch as well as many of the other lookouts.

The sea was smooth with a light swell that made small objects easily visible; drift wood was frequently reported floating on the water. The weather was fair, with a light haze around the horizon reducing visibility to six or eight miles. There were no whitecaps.

At 11:05 a.m., this seeming tranquility was shattered by an explosion that felt to Christy like "a dull heavy thud." A column of water was thrown up along the port side of the ship adjacent to the engine room. As the salt spray settled, the *San Diego* took an immediate ten degree list.

The enemy had struck.

The *San Diego* displaying night illumination. (Courtesy of the Naval Historical Center.)

U-156 (Courtesy of WZ Bilddienst.)

CHAPTER 3
U-156

Although the United States entered the Great War on April 6, 1917, more than a year passed before Germany sent its first undersea raider to American shores. During that time marauding U-boats harrassed merchant convoys in the frigid, often stormy northern waters, sinking freighters, tankers, and sailing vessels along the Great Circle route between England and Canada. The "frantic Atlantic" was the bane of sailors and the merchant marine.

However, the military actions at sea did not touch directly upon American citizens. The battle in the North Atlantic was as much "over there" as the fighting in the trenches.

When the German High Command conceived the plan to attack the American coast, its intentions were threefold: to sink vessels, disrupt shipping, and demoralize the civilian population. Vessels were sunk, shipping was disrupted, but the civilians were not demoralized: they were incensed. Anti-German sentiment rose even higher than before, like the hackles of a maddened dog.

During the six months that U-boat warfare was waged off the coast of the United States and Canada, nearly a hundred vessels were destroyed by gunfire, explosives, mines, and torpedoes, often within territorial waters. One bold U-boat even lobbed gun shells onto New England beaches. Everything from wooden sailing schooners to steel hulled merchantmen to decaying fishing smacks were sent to the bottom, along with their cargoes—and sometimes with their crews.

U-boats first proved their efficacy within weeks of Germany's declaration of war, in 1914, when Kapitan Otto Hersing, captain of the *U-21*, fired a single torpedo into the hull of the HMS *Pathfinder*; the light cruiser went down quickly, with great loss of life. Shortly thereafter, Kapitanleutnant Otto Weddigan, *U-29*, torpedoed and sank in quick succession three British armored cruisers: *Aboukir*, *Hogue*, and *Cressy*. After that came the attack on Allied merchant shipping, the sinking of the *Lusitania* in 1915, and the decimation of the cargo carriers fighting valiantly to transport much-needed food and war materiel to starving nations.

The U.S. Navy slowly—some say ponderously—organized itself for defense against the dreaded underwater juggernauts. At first, most of the fleet was kept in home waters as a floating shield against foreign aggression, an effort largely wasted since the anticipated invasion did not take place. The refusal to meet the threat where it originated allowed U-boats to escape with relative ease through the narrow North Sea channel into the open ocean, where they could wreak havoc among merchantmen.

Eventually, the Navy shifted the bulk of its warships to the European theater: to attack at the head instead of waiting to nip at the heels. The American coast was protected mostly by leftovers, or ships too old or underarmed for war. Vessels of all kinds were transferred to the Navy from other services, and the home guard became a hodgepodge of armed tugs, lightships, Coast Guard cutters, and the still-building 110-foot wooden submarine chasers.

Embarking on the same tactic of attacking the head, Germany finally began a program of U-boat warfare off the American east coast in the late spring of 1918; it was important to cut off reinforcements and the flow of supplies. The United States was unprepared for the onslaught, but not unwarned.

Vice Admiral W.S. Sims, USN, working with the British Admiralty from his office in London, maintained highly accurate intelligence on the whereabouts of all German U-boats. He sent a warning to the Navy Department on April 30: "As there are only seven cruiser submarines built, we are able to keep very close track of these ships.... At the moment the only one that might cross the ocean is the one now coming out of the North Sea."

The U-boat to which he referred was the *U-151*, a *Deutschland* class submarine designed as an underwater freighter. These U-boats were 213 feet in length, 29 feet abeam, and could haul as much as 1,000 tons of cargo. They were built as Kiel, Germany, by Germaniawerft. All seven were completed between 1916 and 1917. Originally, they were unarmed, but after the government took them over they were fitted with two 15-cm (5.9-inch) deck guns and two 8.8-cm (3.5-inch) antiaircraft guns. The *U-156* was incomplete at the time of takeover, allowing for easy installation of two 18-inch below-water torpedo tubes in the bow.

The *U-151* announced its presence off American shores on May 25 by sinking the schooners *Hattie Dunn*, *Hauppauge*, and *Edna*. During the month of June the *U-151* laid mines, and bombed, gunned, and torpedoed twenty more ships in an area that stretched from Virginia to New Jersey.

As the *U-151* rotated home, food, fuel, and ammunition expended, Admiral Sims was privy to an intercepted radio message: "*U-156* putting out, passed the sound in the night of 18 June. U. CR.U." (U-Cruiser Unit.) Sims sent an urgent cable: "Second cruiser submarine at sea. At present off west coast of Ireland. Her field of operations not yet known. Cannot reach longitude of Nantucket before July fifteenth. Shall keep department informed."

The movements of Kapitanleutnant Richard Feldt, captain of the *U-156*, were well known to Allied intelligence, and Sims' suspicions were correct in that the U-boat was on its way across the Atlantic to follow up the devastation inflicted by its sister ship. On June 26, the U-156 torpedoed and sank the British steamer *Tortuguero* some 200 miles northwest of the coast of Ireland.

On July 5, when the *U-156* reached the halfway point of its transatlantic crossing, Feldt fought a battle that can be viewed in hindsight as an amazing presage of poetic justice. The U.S. Navy mine carrier *Lake Bridge* had recently carried a shipment of infernal devices for deployment in the North Sea mine barrage. She was returning to the States in ballast when she was spotted by lookouts on the *U-156*.

The U-boat was disguised with a false funnel to make it look like a merchantman. Feldt tried to steal up on the steamship, but was thwarted by overconfidence. He opened up with his deck guns from a distance of six miles, and managed to land shots from the first salvo close enough to splatter the *Lake Bridge's* hull with shrapnel.

Being armed, the *Lake Bridge* fought back furiously in a running gun battle that lasted thirty minutes. Neither ship nor U-boat was hit. Finally, under cover of a smoke screen, the *Lake Bridge* shoveled coal under her boilers and steamed off at a speed the U-boat's diesel engines could not match. The *Lake Bridge* made it safely to Hampton Road, and went on to carry two more loads of mines across the ocean for the North Sea mine barrage.

The cruiser U-boat's deficiency of speed (12 knots surfaced, 5 knots submerged) forced Feldt to rely on stealth, and to attack only slow-moving vessels. Two days later, still far at sea, he stopped the Norwegian bark *Marosa*, allowed the captain and crew to get away in lifeboats, then laid demolition charges along the sailing ship's hull, and blew her beams out. The men from the *Marosa* were picked up two days later by another Norwegian bark, the *Sorkness*.

U-156 repeated the *Marosa* scenario on yet another Norwegian bark, the *Manx King*. Her men were picked up a day and a half later by the British steamer *Anchisis*. It was not until the ship reached port on July 12 that word got out about the incident.

The U-boat was not spotted again until July 17, this time by the aging troop transport USS *Harrisburg*. The location was southeast of Cape Cod. After remaining in full view for ten minutes, the U-boat strategically withdrew underwater without offering a fight.

Between these two sightings Feldt managed to sneak close to the south shore of Long Island and deposit the mines with which the U-boat was armed. The secret menace lay in wait for some unsuspecting ship to discover.

On July 21, Feldt engaged a tug and her four tows only three miles off Orleans, Cape Cod. For an hour and a half the *U-156* pummeled the *Perth Amboy* with gun shot, eventually setting her ablaze and sinking her barges. People on shore witnessed the one-sided engagement, but were unable to intercede. Seaplanes from nearby Chatham Naval Air Station managed to stave off the attack by dropping depth charges; if the bombs had not been duds they might have sunk the U-boat. Antiaircraft guns returned fire. The planes finally forced Feldt to call off the attack. The crew of the *Perth Amboy*, some injured by shrapnel, rowed ashore, as did the barge crews; the tug was later salvaged.

The next day found the *U-156* off the coast of Maine. Feldt stopped the fishing smack *Robert and Richard*, ordered her men into their dories, and had a bomb placed aboard the hapless vessel. The twenty-three-man crew of the *Robert and Richard* rowed to shore, but not after enduring several harrowing days at sea.

The *U-156* worked its way north toward Canada; for some inexplicable reason it lay low for ten days. In the mean time, however, the *U-140* reached the Virginia coast and began its own war against the merchant marine. Then, on August 2, the *U-156* came out of hiding with a vengeance, and devastated the Grand Banks fishing fleet in the Gulf of Maine.

The first of the next group of vessels sent to Davy Jones' locker was the Canadian schooner *Dornfontein*. Feldt ordered Captain Charles Dagwell and his nine-man crew to come aboard the U-boat. They were held prisoners while the *Dornfontein* was looted and burned. The German sailors bragged about their exploits, claiming responsibility for laying the mine that sank the *San Diego*. After five hours the prisioners were released and told to row for shore, which they made in about three hours.

One after another Feldt ripped through the helpless boats. In the course of a week he sank nearly a dozen, including the 4,868 ton Canadian tanker *Luz Blanca*. After a three day hiatus he torpedoed the British steamship *Pennistone*. Six days later he bombed the Norwegian steamship *San Jose*.

During a six day foray starting August 20, he sank thirteen more fishing boats and one small steam trawler, the *Triumph*, off Canso, Nova Scotia. Feldt did not sink the *Triumph* immediately; instead, he placed a prize crew aboard and used the steamer as a surface raider to lure other ships to their deaths, while keeping his U-boad out of sight.

Feldt accounted for thirty-six vessels during the two month war patrol of the *U-156*; only two were salvaged. His record for tonnage sunk was 33,582, the largest and most important of which was the *San Diego*.

Feldt also made a remarkable record in the scant number of lives lost; twelve in the sinking of the *Tortuguero*, two on the *Luz Blanca* from shell fire, two on the *Penistone* from the torpedo blast, and those on the *San Diego*. Unless the attacked vessel put up a fight, Feldt allowed the crew to escape in their lifeboats before sinking their ship. Without a doubt, many of the seamen suffered greatly during the days rowing or sailing to safety, but at least they were all hardy enough, and close enough to shore, to make the open-boat voyage.

The *U-156* left American waters after sinking the fishing schooner *Gloaming* on August 26. Except for those already mentioned, only three more U-boats made war patrols off the American states and provinces.

Top: The passageway and rail leading to the mine laying tube. Middle and bottom: The forward torpedo room and tubes of a mine laying U-boat. The *U-156* had only two torpedo tubes. (All courtesy of the National Archives.)

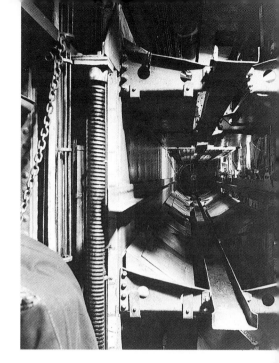

Top and middle: Two views of the tiny chariot bridge of a U-boat. (Courtesy of the National Archives.) Bottom: A U-boat's deck gun. (Courtesy of the Submarine Force Library and Museum.)

While the collective damage they inflicted was great, and the confusion enormous, the German assault began too late and with too few U-boats to prove effective to the outcome of the war; the United Stated was already in full gear. If either country had taken a firm initiative earlier, the story related above would have been drastically different.

On the way home the *U-156* tangled briefly with the USS *West Haven*, a 12,191 ton armed freighter assigned to the Naval Overseas Transportation Service. She was westbound from Bordeaux to New York when her path crossed that of the U-boat. The date was August 31. The two ships exchanged gun fire for about twenty minutes before the *U-156* submerged, and was not seen again by friend or foe. (Although the *West Heaven* survived this U-boat encounter, she was sunk by the *U-402* in World War Two.)

Records of the final days of the *U-156* are understandably sketchy. The last message Feldt transmitted was received by the U-Cruiser Unit on September 24: "Shall pass new English Barrage area on Route 1660 on 25 September after 2014h. Can reach Skagen Reef Light-ship with greatest speed by the evening of 27 September at the earliest. Where can rendevous take place?"

U-Cruiser Unit responded with the following instructions: "Follow planned route only by day and in calm weather. Note carefully large balls of glass which carry mines. Radio entry by Skagen south of new barrage area. Rendevous will then be radioed."

The *U-156* did not make it through the North Sea mine barrage. Perhaps she was sunk by a mine carried by the mine carrier USS *Lake Bridge*. No one will ever know.

Kapitanleutnant Richard Feldt went down with his ship. The log and all the records of the *U-156* were lost, clouding its activities with mystery. Twenty-one survivors reportedly reached the Norwegian coast, where they hid until the signing of the Armistice. The short career of a U-boat that started with such notoriety, ended in anonymity.

A dramatized rendition of the sinking of the *San Diego*. (Painting by Francis Muller, 1920, courtesy of the Naval Historical Center.)

CHAPTER 4
SINKING

At 11:05 a.m., a tremendous explosion stove in the outer bulkhead of the *San Diego's* port engine room. Two men were killed instantly by the blast: Frazier Thomas, Machinist's Mate second class, and James Rochet, Engineman first class. Only moments before, Thomas Davis, Engineman second class, had closed himself in the port shaft alley; he was on his way to oil the shaft bearings.

Lieutenant J.P. Millon was standing watch in the watertight door space between the port and starboard engine rooms. The concussion blew him into the starboard engine room, right over the man working the starboard throttle. Robert Hawthorne, at the port throttle, was knocked from his station into the center line bulkhead.

The forward end of the engine was blown clear off its base, and displaced inward. Within seconds a great inrush of coal-saturated water spurted into the port engine room. Severed steam lines hissed raucously, spewing superheated water into the compartment in thick, scalding clouds. The bulkhead separating the port engine room from Number 8 fireroom was buckled, and the door was blown inboard and twisted sideways. The engine stopped immediately.

On the bridge, Officer of the Deck Lieutenant F.S. Irby "felt a distinct jar, the bridge vibrating considerably." He immediately gave the order to close watertight doors. The quartermaster sounded the warning siren. Irby told the bugler to sound submarine defense (general quarters), and sent a messenger to the radio room.

Above, in the exposed wheelhouse, Captain Christy rushed to the engine order telegraphs and rang for full speed ahead on both engines, at the same time calling for hard right rudder. The port engine annunciator did not answer. The ship took an immediate ten degree list.

Lieutenant Commander Gerard Bradford, the Executive Officer, left the bridge to ascertain the extent of the damage. The first sign of trouble he saw was in the dynamo room, where water was leaking up from flooded compartments below. On the port side one machine was operating with its armature half in the water. The ventilating ducts were pouring a continous stream of water into the room, as were the voice tubes in the central station. Bradford asked Chief Electrician Frank Boot to divert power to the boat cranes.

Ensign John Hildman rushed to the depth charge racks on the stern and cast off the auxiliary lashings in preparation for word from the bridge to drop the 300 pound charges.

From his station in the charthouse, Lieutenant Frank Kutz raced to the forecastle and gave instructions to have the life rafts lowered to the deck. As soon as he was sure that his orders were being carried out, he dropped down to the berth deck to check on the closure of hatches and ports. He ran into Bradford coming out of the dynamo room dripping with water, and was told to make a report to the captain on the bridge.

With water lapping over the supports of the switch board, Boot was pulling power from the lights in order to keep electricity flowing to the boat cranes. Fuses continually blew out as short circuits developed from submerged equipment. Lights blinked out all over the ship.

Lieutenant V.R. Greiff was the ship's radio officer. All attempts to send an SOS were frustrated. The main radio set was knocked out of commission due to the blast interrupting the power leads: the motor-generators were located on the port side of the berth deck. Since the ship was not equipped with emergency storage batteries, Greiff switched the transmitter to a 110-volt outlet in the radio room, but by the time he made the connections the power source failed.

Irby told the quartermaster to call the engine room on the telephone or voice tube; he received no answer from below. The navigator on the bridge, Lieutenant R.J. Carstarphen, leaped down the stairs to the chartroom in order to get their latest position. Christy climbed down from the wheelhouse to the bridge, and relieved Irby so the lieutenant could take his submarine defense station at the starboard 6-inch gun. The list to port was too great for the elevation mechanism: the barrel would not point at the horizon. Although it was impossible to direct an attack against an enemy that remained unseen, Irby pulled the percussion firing pin anyway and shot a round into space; it was the first time in the *San Diego's* career that her guns were fired in anger.

With the ocean gushing into the inoperative port engine room, Lieutenant Millon gave orders to close the watertight door separating it from the starboard engine room. However, this door was jammed. In only half a minute the port engine room was full to overflowing. Millon found the door between the starboard engine room and Number 7 fire room operating smoothly, so he stepped through and closed it behind him. But when he opened the door between fire rooms seven and eight, he was met with inrushing water. He closed the door, then gave the order to open the feed stops in order to keep the water in the boilers and prevent an explosion. He then told the men to abandon the fire room.

If Thomas Davis was still alive, he was trapped inside the shaft alley.

The engine room was by this time a cacophany of steam escaping through tortured pipes. Despite Captain Christy's order for full speed ahead, the annunciator bell in the engine room rang for all stop, then, a few seconds later, full speed astern: the chain linkage had been knocked askew by the blast. Millon and the only other man still working the engine, George Stockton, did their best to comply with the erroneous order. Torrents of

Top: A sailor makes adjustments to the *San Diego's* engine. On the bulkhead behind him is the engine room annunciator. (Courtesy of the Naval Historical Center.) Middle: An annunciator on the wreck is thickly overgrown with anemones. Bottom: An annunciator that was recovered, restored, and photographed by John Lachenmeyer.

water poured through the unsealed center-line hatch. When the water stood four feet deep over the floor plates, Millon and Stockton hurried to carry out last minute shut down procedures: Millon alerting the bridge on the annunciator that he was stopping the engine, while Stockton closed the proper valves.

The two men climbed the catwalks and escaped into the berth deck at a level above the waterline and the armored deck; the door leading into the passageway had been carried away by the blast. Tom Humphrey reported that the deck plates in the machine shop were buckled as well. Lieutenant C.J. Collins, Engineer Officer, appeared at that moment to inspect for damages before reporting to the captain. It was his opinion that, since the top of the evaporator room was the deck of the machine shop, the door between the port engine room and the evaporator room must have carried away as well.

Topside, Lieutenant Irby jumped from the casemate to the gun deck, but when he tried to reach the berth deck to check the closure of the watertight doors, he found the engineer's log room filled to the top with sea water. Because the ship was still rolling to port, inrushing water flooded the berth deck. Another watertight door was broached by the pressure, allowing the overflow to cascade through the armored hatch into the

This model of the USS *Pennsylvania* (ACR-4) is on display at the Washington Navy Yard, Washington, DC. The helm just forward of the mast is where Captain Christy stood as he conned the *San Diego* during her final moments. A similar model of the *California* is on display at the San Diego Maritime Museum.

starboard engine room: It was now being flooded from above as well as from below.

Captain Christy remained at the conn, digesting damage reports. The turn to the north steered the *San Diego* toward shoal water; she was only ten miles from shore. But without motive power the ship slowly lost headway.

Knowing that the barely moving ship could not get out of range of her own depth charges without suffering damage, Ensign Hildman secured the depth charge racks. He tied the safety forks with line so they could not be released by sudden jars.

The starboard 6-inch gun crews fired their guns at what they thought was a periscope. Shells winged across the ocean at anything that remotely resembled a German submarine. The 3-inch guns joined the fracas as well. On the port side the water was lapping at the lower part of the gun shutters.

With the main radio off line, Lieutenant Greiff carried the field set to the badly tilted boat deck, connected the antenna, and tried to start the backup gasoline engine. The trip up the ladder had spilled gasoline out of the carburetor and flushed the oil out of the bearings; the engine would not start. The ship was completely without power.

Captain Christy quietly gave the order to abandon ship.

The word passed quickly through the echelons. Men poured out of darkened corridors and below-deck compartments in answer to the call, swarming onto the main deck in order to don life preservers and canteens. Some of them struggled to launch boats by hand. There was no confusion. The abandonment might have been just another drill.

Lieutenant Irby descended to the port-side gun deck, carrying an armful of life belts. The gun crew was still at its station, firing the gun although the muzzle was nearly underwater. Irby had to pull some of the men away from the breech and order them overboard. The men took the life belts and simply stepped through the gun port into the water.

So quickly was the *San Diego* rolling that by the time Irby climbed the ladder to the boat deck it was even with the water. The other gun crews left their posts only after they could no longer fire their weapons. Counting all the guns, about forty shots had been fired.

Lieutenant Grieff's duties called for him to throw the code books overboard. He left the malfunctioning radio set and headed directly for his cabin safe where most of the secret information was stored. Water had risen so high in the ward room, and the angle of the ship was so severe, that he could not open the heavy steel door; he left them. One set of operational codes, in lead-covered books, was tossed into the sea; another set was saved.

As the ship heeled sharply all manner of loose gear slid down the decks and splashed into the water. Guy wires snapped, and one by one the stacks broke off and crashed into the sea; one of the stacks crushed the port motor sailor.

Acting Pay Clerk J.D. Gagan removed the currency from his safe: $20,305 of government funds, and $720.11 of sailor's safe keeping money. Left behind was $5 in gold, some $700 in silver, and approximately $38,259 in currency.

The ship was rolling even faster now. Two whaleboats, a dinghy, a wherry, and two punts were launched off the steeply canted deck. Life rafts were cut free and slid into the water. Anything that would float was tossed overboard, including a hundred kapok mattresses, fifty wooden mess tables, loose hammocks, and a pile of lumber carried as deck cargo.

Almost leisurely, over a thousand men leaped into the water and swam away; they clambered onto whatever floating jetsam was available. D. Easdale, ship's carpenter, walked overboard when the ship went out from under him. Ensign Hildman merely floated off the cabin hatch. The *San Diego* was nearly on her beam ends.

Machinist's Mate Andrew Munsen was struck in the head by one of the rafts, and was killed. Engine Man Clyde Blane drowned. Seaman Paul Harris was the last lookout to start down from the crow's nest in the forward mast; he was apparently still inside the cage when the mast hit the water, and was dragged down with the ship.

Only two men were left aboard. Bradford went over the port side. Captain Christy, in the age-old tradition of the sea, was the last man to leave his ship. He jumped from the nearly vertical bridge and climbed down a ladder to the boat deck, then slid down a line to the protrusion of the still-dripping armor belt. As the *San Diego* continued to capsize, Christy walked across her freshly cleaned bottom like a logger balancing on a rolling log. He kept pace with the ship rolling under him, dropped from the armor belt to the starboard bilge keel, then jumped to the docking keel. He hesitated while surveying his ship, to make sure no one else was aboard. Then, saluting his command, he jumped eight feet into the water.

Men in the rafts cheered as the captain and the executive officer left the ship. Christy was without a life jacket; he swam a short distance from the settling cruiser, and was picked up by a motorized whaleboat.

It took another five minutes for the ship to turn completely upside down, in a symmetrical position with the keel inclined about ten degrees to the horizontal, the forward end elevated. The *San Diego* sank slowly by the stern, without appreciable suction. Total time from first explosion to last sight of the inverted hull was twenty minutes. Remarkably, only six men died.

In the water were 1,177 survivors: 49 officers, 9 midshipmen, 63 marines, and 1,056 enlisted men. The men were in excellent spirits. They sang "The Star Spangled Banner" and "My Country 'Tis of Thee." When the U.S. ensign was hoisted on the sail boat, they cheered the stars and stripes. The men's love of their ship did not end with her sinking; they were true patriots to their service and to their country.

An earlier photo of the *California*, taken when the guns were still in place. (Courtesy of the Naval Historical Center.) View the picture vertically, and imagine Captain Christy walking horizontally from the bridge to the hull as the ship rolled onto her side.

Captain Christy ordered Lieutenant Clarkson Bright to take the dinghy to shore to alert the Navy of the *San Diego's* plight, and to request rescue vessels to pick up the men. Two hours later Bright and twenty sailors landed through the surf at Point o' Woods. Since Coast Guard Station No. 82 was nearby, authorities of the Third Naval District soon learned of the disaster. The first Navy vessel to leave for the scene of the disaster was the destroyer *Perkins*, followed by the torpedo boat destroyer *Preble* and the torpedo boat *Shubrick*. Soon a small flotilla of boats was underway: Coast Guard cutters *S.P. 740*, *S.P. 966*, *S.P. 251*, and submarine chasers *SC-55*, *SC-56*, and *SC-59*. Later, more boats joined the hunt.

Bright and the sailors were treated handsomely by local residents, who gave them food, clothing, and drink. That afternoon they were taken to New York by automobile.

Meanwhile, Captain Christy was not sitting idle. He had the whaleboat's mast stepped, and hoisted sail. The whaleboat then carved a circuitous course through the survivors, picking up the sick and injured. Two steamships were sighted to the south. The whaleboat tacked toward the smoke and soon intercepted the *Malden*, Captain H.R. Brown, and the *Bussum*, Captain Brewer. Despite the possibility of a U-Boat's presence in the vicitity, both captains unhesitatingly came to the rescue. Shortly thereafter, the *F.P. Jones*, Captain Dodge, hove into view and joined the operation.

Every man of the *San Diego's* crew was picked up before district headquarters at Sayville knew of the casualty. The three vessels searched the area until 3 p.m., then proceeded to New York with the survivors: 708 on the *Bussum*, 370 on the *Malden*, and 78 on the *F.P. Jones*.

Representatives of the American Red Cross met the beleaguered crew men with comfort kits, sweaters, socks, pajamas, watch caps, and 1,200 blankets. Since the men lost everything they owned except what they were wearing, the gifts were sorely needed. Later, the crew was given quarters aboard the USS *Maui*, a troop transport.

At the same time, the hunt was on for the dastardly U-boat that sank the *San Diego*. No one reported seeing a submarine, either its wake or its periscope. Yet, because the damage occurred around frame #78, which was abaft the point of the ship's greatest beam, it seemed unlikely that a mine could have caused the explosion.

It was five o'clock before the *Preble* arrived at the approximate position of the sinking. *Preble* reported "considerable wreckage floating around, in center of which there was what appeared to be a smokestack at an angle of 45 degrss, and perhaps two or three feet out of the water." She picked up five Carley Life Rafts and a punt.

By this time the sea was a beehive of activity, with sub chasers and mine sweepers crashing through the waves searching for signs of the enemy U-boat. Seaplanes (or, as they were called at the time, hydroaeroplanes, or hydroplanes) buzzed in the air. Seaplane No. 7 spotted what it thought was a U-boat; it flew close enough to the *Shubrick* for the pilot to shout a verbal warning, then flew off to make an attack.

The plane dropped an aerial torpedo over the suspected U-boat, which was submerged, but the torpedo failed to explode. The plane then returned to the slower-moving torpedo boat, and dropped a box containing a message at her stern. Seaman Second Class William Queenan jumped off the *Shubrick's* fantail to recover the box. The message read "Submarine—follow us."

Shubrick made off at full steam; she signaled three of the sub chasers to follow. General quarters was sounded and the ship prepared depth charges. All the men on the torpedo boat saw what looked like the wake of a submarine. The *Shubrick* made a run on the spot and dropped a depth charge which did not detonate. She circled, and dropped another depth charge; this one went off, as did two more shot from the Y-gun of the *SC-56*. The men then reported seeing an object very much like a periscope rise momentarily seven feet above the surface, then disappear. *Shubrick* circled again and discharged her last depth charge right on target. She then fired her port battery at the location.

Perkins arrived, saw the spot where bubbles were frothing the surface of the sea and where oil stained the blue ocean with rainbow colors. She dropped a pattern of four depth charges, none of which exploded.

SC-55 fired her Y-guns. The *Preble* steamed into the fracas and fired her Y-guns as soon as she came abreast of the still-visible wake. Both charges detonated. She circled three times, each time dropping another deadly can of explosives on the rising bubbles. She also fired her deck guns at what appeared to be a periscope.

The progression of bubbles seemed to be moving in a straight line, so the *Perkins* chased the apparently moving object with shells from her forward guns. When the bubbles maintained a stationary position, the *Perkins* let go her anchor and dragged over the spot, hanging on a submerged object for a while.

Then a pair of mine sweepers came in and swept the sea. They caught the object in their wire, and held onto it all night long. Five more mine sweepers worked their wires. Two groups of sub chasers prowled in the vicinity; they picked up what they thought was a moving submerged object, but lost track of it.

In the morning, two more mine sweepers formed a loop over the object caught by the first two, the idea being to secure the object in case it turned out to be a U-boat. They maintained their position for the rest of the day. Other mine sweepers picked up six mines, all of which were exploded by gun fire. With enemy mines now known to exist, sub chasers raced back and forth to warn ships of the danger.

The whole area was a mass of floating wreckage, much of which resembled floating mines. Cautious skippers did not draw near anything without shooting at it first. In retrospect these pot shots may seem wasteful, but in light of the situation as it was perceived the tactic was sound. No one knew what manner of infernal machines the Germans had left. It was only after the objects fired upon failed to explode that they were approached and identified: paint drums, pickle kegs, coffee tins, and water breakers.

By this time, of course, the *U-156* was several hundred miles away, attacking the fishing fleet between Maine and Nova Scotia.

The *San Diego's* crew stayed on the *Maui* for a week, waiting for orders. On July 26, all officers and enlisted men left for Pelham Bay Training Camp, where they were quartered in the 8th Regiment Barracks. Many men received seven days leave. With the San Diego gone, reassignment to other vessels was the only option available. This was a sad undertaking for the men, for most of them had served continuous duty on the San Diego for the previous three years.

Long-time friendships were bound to be separated. Starting August 17, the men began being transferred to Receiving Ships for further assignment. August 24 saw more transfers, as did August 26. On September 1 and 2 some men left for the U.S. Navy Training Center.

Captain Christy received his orders on September 3: he was given command of a battleship, the USS *Wyoming* (BB-32), then in Scapa Flow. He reported aboard on September 25. The men were mustered on deck on September 29 for the formal announcement. At 1:00 p.m., Captain Christy officially relieved Captain H. Wiley of his command. The *Wyoming's* deck log reveals very little fanfare, for it then goes on to note that at "1:15 received 30,240 pounds of potatoes from *Volana*."

The Secretary of the Navy was more kind. Christy was given a commendation for the "admirable action of the commanding officer of the U.S.S. *San Diego* at the time of the sinking of that vessel by an enemy submarine."

Ten days after the armistice the *Wyoming*, along with the *New York*, *Texas*, *Arkansas*, and Britain's Grand Fleet, escorted the German High Seas Fleet into internment at Scapa Flow. In December, the *Wyoming* rendevoused with the transport *George Washington*, which was carrying President Woodrow Wilson to the Paris Peace Conference. Christy continued his distinguished service in the Navy, eventually rising to the rank of Rear Admiral.

For the rest of the *San Diego's* crew came sea duty. The Pelham Bay muster roll grew smaller each day. At morning muster over the next two weeks the names of ships rang out like a littany of the Naval Register: *Pleiades*, *Henry W. Mallory*, *Galveston*, *Broad Arrow*, *Pocahontas*, *Sylvan Arrow*, *Samarinda*, *Western Front*, *Powhatan*, *Tjisondari*, *Sherman*, *Pasadina*, *Julia Luckenbach*, *Edward Luckenbach*, *Manchuria*, *Tacoma*, *Pueblo*, *Standard Arrow*, *Burtenzord*, *Aniwa*, *Louisville*, *Fairfax*, *Columbia*, *Colgoa*, and *Ice King*.

By September 15 they were all gone, the last detached to Receiving Ships at New York, Philadelphia, and Washington, DC. To the far flung corners of the world the valiant men of the *San Diego* sailed. There were battles to be fought, a war to be won, freedom to be defended. The men of the *San Diego* went on to perform their patriotic duty in whatever sea they were needed, on whatever ship they could serve. They were sailors in the grand tradition of the fleet.

The *San Diego* was gone, but not her fighting spirit: the men would carry that with them wherever they went. With such honor the United States Navy is imbued; with such pride it is fraught.

Harley Christy as a captain (left) and, later, as an admiral. (Courtesy of the National Archives.)

Comparative views, above as the *California* (courtesy of the Naval Photographic Center) and below as the *San Diego* (courtesy of the National Archives.) Notice the structural differences, particularly the forward mast, anchor placement, and bow gilt. The flag in the upper picture has forty-six stars. (Oklahoma joined the Union on November 16, 1907, New Mexico and Arizona in 1912.)

CHAPTER 5
SALVAGE

The location of the object that appeared to be a smokestack angled at forty-five degrees was lost. The wreck that the mine sweepers were holding turned out to be a coal barge; a diver from the *Perkins* identified it without question the next morning. But three miles to the south the *S.P. 427* located the remains of the *San Diego*.

Lieutenant Commander Edward Keen, commanding officer of the local mine sweeping division, was aboard the *S.P. 427* while the area was being swept for mines and while the search was going on for the sunken cruiser. After mine sweeping operations had been secured for the day, and the six discovered mines were either sunk or destroyed, Keen decided to do a little investigation on his own. He located an object from which bubbles and oil were escaping. When he dropped a sounding lead on it, the bottom of the lead came up tinged with gray paint reminiscent of that used to cover the bottom of Navy ships' hulls. He marked the position with a second class nun buoy.

On July 20 USS *Bagley*, a torpedo boat, arrived on site with a Navy diving party. The *Bagley's* narrow beam caused her to roll excessively in the heavy seas; diving from her was impractical. Instead, the motor sailor from the *Perkins* was used as a platform; it was held some twenty feet off the *Bagley*, where the pumps were kept. A ladder was rigged over the stern of the motor sailor enabling William Williamson to go down in his hard hat rig.

Williamson made a nine minute dive starting at 6:00 p.m. He landed on the hull of a huge vessel, then slid off the side. He saw enough to identify the wreck as that of the *San Diego*. He returned topside to make his report, then went down again six minutes later.

During an eighteen minute dive he pulled himself along the bilge keel for about two hundred feet; a strong tide was running. He reported the wreck to be lying in 18 fathoms (108 feet), bottom up, with the smoke pipes bent to one side. The bow was slightly higher than the stern, the least depth being 8.5 fathoms (51 feet) at the bow and 10 fathoms (60 feet) at the stern. He found a jagged hole five feet in diameter on the port side abreast the number four stack, and about twelve feet below the water line of the sunken ship.

Williamson also found numerous holes, dents, and sheared rivets: the results of a plethora of depth charges dropped on the *San Diego* in the mistaken belief that she was a submarine. Bubbles of air streamed out of the hull like water through a sieve. He was finally swept off the hull by the current.

Merritt & Chapman Derrick and Wrecking Company was hired to conduct a second examination of the wreck. On July 27, working off the USS *Passaic*, Joseph Rouleau and John Gardner each made a single dive.

The bow of the *San Diego* pointed approximately north; the hull was inclined about twenty degrees from vertical with the starboard side uppermost. They determined that no more than 36 feet of water separated the starboard bilge keel from the surface. There is no explanation for the discrepancy in depth readings between the two divers' reports, but the latter depth was proven valid.

In addition to the engine room hole, they found the adjacent section of the hull smashed in for about twenty feet between the lower edge of the armor plate and the bilge keel, and the plates torn apart. Loose rivets and shims lay all over the bottom. Forward of midships, from bilge keel to bridge deck, the plates were bent in and butts and seams torn apart. One 3-inch gun was broken from its mount.

They walked under the main deck from the starboard side. Both turrets were in place with the tops resting on the sand; the barrels were trained fore and aft. Masts and smokestacks lay twisted sideways protruding to starboard. Air was coming out of the hull from stem to stern. As the wreck continued to lose buoyancy it undoubtedly would settle down deeper, crushing the superstructure.

Most importantly, however, Gardner found an unexploded depth charge lying on the hull adjacent to the boiler rooms. That terminated all diving operations. If it were jarred free and rolled off the wreck, it might go off as soon as it reached its preset depth setting.

On July 27, an obstruction gas buoy was established over the wreck site, to denote it as a hazard to navigation.

Naval Constructor W.N. Davis recommended that, although it might

Left: A drawing of the wreck made after the initial survey. The roll of the ship as she capsized, plus the angle of tilt after coming to rest on the bottom, explains why much of the *San Diego's* loose parts should have slid to the port side. Right: A depth recorder printout of the *San Diego*.

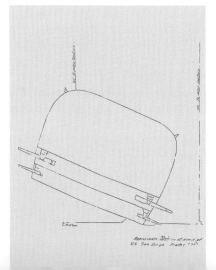
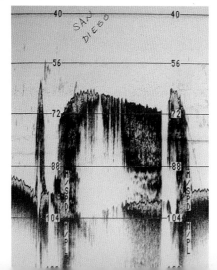

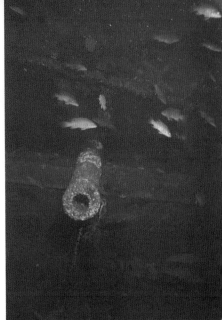

3-inch guns protruding from their ports.

be possible to recover some of the 6-inch and smaller guns, the work would probably prove so expensive and uncertain because of the wreck's upside down position, that the undertaking was not feasible. Total hull salvage was out of the question because of the profusion of leaks and loosened rivets.

Consequently, on August 5, an endorsement from the Bureau of Construction and Repair stated that no attempt would be made to salvage the *San Diego*, or to recovery any of her guns. Her name was stricken from the Navy List on August 26, 1918.

The USS *Resolute* took soundings on the wreck on October 15, and found that it had settled two feet since July 26. A year later the wreck had collapsed enough that it was no longer a menace to navigation. The superstructure on which the wreck rested was not designed to withstand the massive deadweight tonnage of the armored steel hull; it was slowly being crushed like a folding accordion.

As currents whipped around the wreck, washing out the sand, it sank even deeper. Subsequently, on October 21, 1919, the buoy was discontinued, and the wreck symbol on the U.S. Coast and Geodetic Survey charts was removed.

The only recovery of the *San Diego's* effects was a paper recovery; and it was largely the newspaper that brought it about. An editorial in the *New York Times* for October 25, 1918 told how the men of the *San Diego* "had taken about $100,000 worth of First Liberty Loan bonds, that they had been without shore leave since the bonds were delivered to them, and that these securities, with all the other property of the crew except the clothes they wore, went down with the ship."

The Federal Treasury refused to "reissue bonds to anybody who cannot give satisfactory proof that the old bonds have been destroyed."

According to law, the Treasury Department was permitted to duplicate "coupon bonds when they are destroyed, but not when lost."

The *New York Times* kept up its editorial tirade, incensing American citizens against such governmental impropriety. Even so, it took legal action combined with the voices of outraged Congressmen and Senators to convince the Treasury Department to reconsider its policy. Chief Water Tender James Poteet filed suit against the very government for which he had been fighting; he had lost $350 worth of bonds when the *San Diego* went down. This was the test case.

Fighting for the rights of its men, the Navy Department stated that "such extensive damage had been caused as to make it certain the Liberty Bonds had been destroyed," and that "no attempt would be made to salvage the *San Diego*," or the bonds contained therein.

Finally, Assistant Secretary of the Treasury, R.C. Leffingwell, was forced to issue a statement which said, in part, "You are advised of the receipt by this office of information from the Navy Department that more than thirty depth bombs were dropped at the spot where said vessel (the *San Diego*) was sunk, and subsequent examination by a diver disclosed great destruction caused thereby."

More than a year after the men made good their escape from the capsizing cruiser, they were reimbursed for the loss of their bonds. Patriotism has its just rewards, but sometimes they are late in coming.

Even then, the *San Diego* was not forgotten. On May 16, 1921, the New York based Saliger Ship Salvage Corporation sent a letter to the Commandant of the Third Naval District stating its desire to salvage the wreck. A brochure submitted with the letter described the patented devices that were part of its salvage system: a burrowing machine (for drawing lifting cables under sunken ships), resilient pontoons, a sub-sea vision scope, a self-fitting sealing patch, an acetylene underwater steel cutting torch, and an inductance process for locating sunken vessels.

According to Saliger's advertising hype three versions of the burrowing machine were available: "one resembles a caterpillar tractor combined with hydraulic sluicing; another resembles a double torpedo with oppositely rotating propellers, and still another resembles a mole. These machines are operated hydraulically or electrically and are controlled and directed by a

Ammunition had a realistic value when it was still usable.

Personal effects. Left: A silver cigarette case whose monogram reads HHLJr. According to the muster role, Herbert H. Law, Jr. was the *San Diego's* yeoman. Middle and right: This gold locket recovered by Jon Hulburt belonged to Ensign Vernon Jannotta.

pilot at and from a distance. The Burrowing Machine is not a dream. It is a proven out practical reality. It has been built and demonstrated, and is ready now for business."

The hard sell was evident in Saliger's call for investors. "Hundreds of wrecks lying at the bottom of the sea are veritable gold mines, each one containing in concentrated form and confined within the space occupied by the wrecked vessel more actual value than is ever produced from the average gold mine."

If Saliger's intentions were sincere, the Navy did not believe so. Asked for input, the Bureau of Construction and Repair replied, "As this Company apparently is selling stock, it is suggested that they may desire this permission more for advertising purposes than with an intention to prosecute salvage operations seriously."

Notwithstanding such observations, the navy's solicitor issued a Legal decision on June 11: "The office of Naval Operations advises that it sees no objection to granting permission to the Saliger Ships Salvage Corporation to salvage the *San Diego* as long as no expense to the Government is involved, and provided that the Government shall have a preferred right to obtain by purchase such material from the vessel as may be desired in case the operations are successful." Furthermore, "the Government assumes no obligations whatsoever in connection with the matter, the Corporation to undertake the work at its own expense and risk."

Certainly, the *San Diego* was a wreck worth salvaging. Besides the hull, which could be cut up, melted down, and sold as scrap iron, there were the ship's guns and appurtenances that might be restorable. Undoubtedly, the 20,000 pounds of beef and 30 tons of dry provisions were no longer edible. Perhaps some of the $30,342.59 worth of clothing was recoverable, or the ship's store's $6,680.22 worth of salable supplies. It is possible that the $225,000 worth of ammunition and ammunition containers could be reused, as well as some 2,750 tons of coal. The Bureau of Supplies and Accounts estimated that between $2 million and $3 million worth of miscellaneous material was lost in the sinking. Undoubtedly, this latter figure included the cost of hull replacement.

Perhaps Saliger's description of sunken gold mines was more truth than trivia. Yet, one wonders why, in all the ship salvage operations performed throughout the world since that time, no other mention of a burrowing machine has ever been made. Nor did Saliger ever carry out its proposed scheme to salvage the *San Diego*.

Perhaps, when the *Brooklyn* was scrapped that same year, Saliger realized that the salvage value of the *San Diego* would be less than the anticipated cost of reclaiming her, which itself was problematical.

By 1930 it became painfully obvious to the Navy that the armored cruisers were not the concentrated gold mines of Saliger's imagination. Six of them went to the ship breakers that year, and another the next. Since the *Memphis* had been wrecked in the Dominican Republic in 1916 by a series of abnormally large tidal waves, and subsequently salvaged for scrap, only one of the Big Ten armored cruisers was still extant, plus the ancient *Rochester*.

The *Rochester* was decommissioned in 1933; she spent the next eight years rusting away at the Olongapo Shipyard, in Cavite, until 1941, when she was scuttled to prevent her capture by Japanese invasionary forces. The *Seattle's* career ended in 1927, from which time she served as a receiving ship at New York until her rust-pocked hull was scrapped in 1946.

They were all gone: dismembered, melted down, or otherwise anatomized. None of their constituent parts remained; if they existed at all it was in the form of raw iron and other atomic particles. All, that is, except for one. For at the bottom of the cold blue sea, maintaining contiguity, lay the intact hull and appurtenances of the last armored cruiser: the USS *San Diego*.

Seasons came and went, the years passed, the decades piled atop one another. In relative obscurity the *San Diego* rested peacefully on her bed of sand. Then, with very little fanfare, the upside down hulk was sold to another New York based salvage firm. On october 29, 1957, Maxter Metals Corporation paid the Navy $1,221.00 "for scrapping only."

Maxter Metals put divers on the *San Diego* in order to conduct a preliminary survey. The plan that evolved called for blasting the wreck apart with dynamite, and salvaging it piecemeal. For six years Maxter Metals kept the *San Diego* as an ace in the hole, waiting for a slump in business, or a dramatic rise in the price of steel. The time was not yet propitious for an all-out salvage operation.

Fortunately for the continued existance of the *San Diego*, Maxter Metals waited too long. In 1961, a group of environmentalists formed the American Littoral Society, whose avowed purpose was to protect the delicate coastline from manmade destruction, and to preserve the tidewaters and offshore regions as viable wildlife habitats. The first effort of the ALS was a fish count, in which amateur naturalists made identification of species and observed behavioral patterns which paid scientists were too busy or underbudgeted to make. The establishment of a quarterly journal called the *Underwater Naturalist* offered a place to record the data.

Although shipwrecks were not originally prescribed as part of this preservation process, the 1962 salvage of the U.S. submarine *G-2* off the connecticut coast, in addition to the removal of several other long-time shipwrecks, spurred concern over the threat of other planned salvage operations which would reduce the number of prime fishing spots which anglers were wont to fish.

The legality of shipwreck salvage was only then becoming a thorny problem. Previously, salvage was encouraged because it not only returned to the stream of commerce valuable resources otherwise forfeited, but because it reduced hazards to navigation at no expense to the government. Additionally, commercial fisherman who frequently lost expensive nets and trawler gear were glad to have such obstructions removed from the sea bed.

On the other hand, party boat operators who made their living on wreck fishing, and private boat owners and spearfishermen who frequently fished the wrecks for sport, were upset over the loss of what they considered prime spawning grounds for tautog, ling, black sea bass, cod, pollock, and porgies. In the Atlantic underwater ecology, each shipwreck is an oasis in an otherwise barren wasteland: an artificial reef in a desert of sand. In fact, the more progressive coastal states were instituting a program of sinking overage vessels and barges full of trash in order to create fish habitats for the future. To citizens and other interested organizations, it did not make sense to sink a ship on one hand while salvaging one on the other.

In relation to the *San Diego*, the situation reached a fever pitch early in 1963 when ALS members stated that Maxter Metals planned to salvage the cruiser that summer. The ALS formed a special Underwater Preserves Committee, which held a wreck conservation meeting on March 6 to discuss the issues at stake. Spearheading the battle was Paul Tzimoulis, chairman of the committee. The objectives were wreck conservation (prevention of the destruction of wrecks), creation of artificial reefs, and the eventual creation of underwater parks.

At the meeting, lectures were offered on such subjects as "Underwater Wreck Destruction," by Tzimoulis; "Biological Importance of Wrecks," by John Clark, president and spokesperson of the ALS; "Wrecks, a Vital Part of Party Boat Industry," by Leroy Pearsall; "Historical Significance of Wrecks," by Hunter Ross; and "Wrecks, the Skindiver's Underwater Parks," by Nixon Griffis.

The news media was invited to the Sandy Hook Marine Laboratory at Highlands, New Jersey, to listen in. Said Tzimoulis, "We've got to get the public aware of this. State and Government people should become interested and start working for us. We must get the conservation idea over." Afterwards, local newspapers and fishing magazines added their weight to the fight for preservation in the form of timely editorials that reached the public most affected by the ongoing salvage programs.

Tzimoulis wrote letters to all the coastal state legislatures, as well as to the U.S. Navy and the Department of the Interior, informing them of the problem. Responses were mixed, ranging from apathy to total support.

While the Navy suggested that "the extensions of the performance bond for the removal of this wreck have been granted by the Defense Surplus Sales Office of Brooklyn, New York. Any further extension is a matter under the cognizance of that office," the Underwater Preserves Committee was taking an alternative tack.

The *San Diego* Fund was instituted, sponsored by the American Littoral Society, in cooperation with the National Party Boat Owners Association and the National Association of Marine Angling Clubs. Participation was sought throughout the fishing and conservation community. Because the Navy suggested that "one means of preserving those wrecks which now exist would be for your society or other interested groups to acquire title to them." Wreck stamps were issued; they could be bought for one dollar. The ALS hoped to raise enough money to purchase the *San Diego* from Maxter Metals; it was prepared to tender an offer of $15,000.

Meanwhile, Tzimoulis continued to exert political pressure. In contradiction of Admiralty Law governing the salvage of shipwrecks, he stated that "it is the feeling of the Littoral Society that it is up to individual States to create laws governing the title and disposal of abandoned wrecks in Territorial waters." He stressed "the preservation of wrecks as desirable fish habitats."

Furthermore, John Clark spoke "of the need for an intensive and complete survey of existing wrecks, beginning in the area between Cape Hatteras and Cape Cod, and eventually extending to all U.S. Territorial waters." Saving the *San Diego* was therefore not an ultimate goal, but a test case in jurisprudence for future shipwreck preservation.

In August, Maxter Metals' contract for salvage rights on the *San Diego* expired. The Defense Department, made aware of the wreck's value as a marine life habitat, stated that it "will not make a decision to re-offer the vessel for sale without prior coordination with the Department of the Interior and any other Department or Agency that evidences an interest in the matter."

The *San Diego* was saved from demolition.

While the ALS claimed a victory in staving off salvage by Maxter Metals until the expiration of its contract, the news media wrote that the salvage firm "signified a willingness to negotiate transfer of its rights to the *San Diego* wreck to the American Littoral Society for conservation purposes."

When this author wrote to Maxter Metals in 1983 asking for clarification of its position with respect to the above-mentioned contradiction, the response stated that ownership "which we possessed long ago, has, many years ago, been relinquished by us. In fact, there is no one today in our office to even remotely fill you in on any details."

In any case, as far as total hull salvage was concerned, the *San Diego* was secure. With newfound awareness, public outcry would undoubtedly shout down any further attempts to dynamite the wreck into oblivion. As a fish haven the value of the wreck—indeed, of all wrecks along the reefless Atlantic bottom lands—is now well understood and appreciated. Paul Tzimoulis and the American Littoral Society orchestrated the education of both an uninformed populace and apathetic political authorities.

While salvors may seem to have been unfairly castigated for prosecuting their business, they likely lost very little, especially in light of the overall savings to the fishing industry. And while there is cause for concern in a democracy when one faction advocates legislation that disadvantages its competitors, the salvors did not deem the issue important enough to complain. Maxter Metals held salvage rights for the *San Diego* for six years without acting, nor did the company rush into blasting operations or put up a fight as its contract drew to a close; its quiet withdrawal suggests professional equanimity. In the final analysis, the trade-off appears to have been fair.

The resultant status of the *San Diego* as a fish haven is purely unofficial: no statutory designation has ever been proclaimed for the sunken cruiser. Since the U.S. Navy formally abandoned the wreck, and struck the ship from the Naval Register, the legality of Navy licensing is suspect. Because the wreck lies in international waters, outside governmental jurisdiction, Admiralty Law retains authority: salvage of the *San Diego's* hull or appurtenances remains unrestricted. Legally, the wreck can still be salvaged in whole or in part by whomever is willing to bear the risk of financial burden.

A state or federal court order prohibiting salvage would affect only U.S. citizens, not foreign salvors: a nebulous predicament which would have no effect in preventing the wreck's ultimate demise. The only law ever passed with respect to the *San Diego* is one that cannot be repealed, and from which there is no protection: natural law. Eventually, all wrecks succumb to the degenerative process of the corrosive environment in which they are moored. The *San Diego's* worst enemy is not the salvor, but nature.

The *San Diego* in dry dock. (Courtesy of the Naval Historical Center.)

While schemes of salvaging armor of questionable value have fallen by the wayside, the *San Diego* has been the object of two further operations that sought to recover metal of less dubious worth: the seventeen tons of manganese bronze (naval, or marine, bronze) that comprised each propeller. The removal of particular items is in keeping with the intent of maintaining the wreck as a marine life sancturay since it does not destroy the overall integrity of the hull.

The first job was led by Ben Manuella, a diver and ship artificer who, along with a small coterie of fellow workers, spent his weekends exploring the shipwrecks that abound in the New Jersey—New York bight. They decided to blend sport with profession, and utilize the skills they had acquired in their occupations.

During the course of a month in the late sixties (Manuella no longer remembers the year), they made several multiday trips to the wreck site on a two-hundred-foot-long yard oiler: a miniature tanker fitted out with a deck house and living quarters, and which carried fuel oil in drums. Each time, the men dragged heavy cables down to the wreck and with burning bars cut away more of the port propeller shaft.

When the propeller broke off and fell to the sand, they rigged it for the lift. The fairlead of a crane was welded to the bow of the yard oiler. They winched the propeller to a point just below the surface, where it was slung during the trip to Staten Island. On the way, a weak roller gave way. The cable hit a sharp edge, sheared, and as the propeller plummeted to the

The *San Diego* propeller. (Photo by Mike de Camp.)

bottom the loose end of the cable whipped over the bow and strafed the winch deck. Fortunately, no one was on the winch deck when the accident occurred.

As far as Manuella knows, the propeller still lies in the sand somewhere between the *San Diego* and Staten Island.

The salvage of the starboard propeller also involved a strange turn of events. The tale has been often told, and always misrepresented. Ed Betts gave the author the inside story.

A partnership was formed between Ed Betts (master diver), George Dyott, an experienced salvage master and the owner of the 77-foot tug *Sailfish*, and Frank Davidson.

Davidson owned the 140-foot barge *Lehigh Valley 401*. In addition to a 20-foot by 20-foot deck house, the barge sported a derrick with a 77-foot A-frame and an 85-foot boom. The *Lehigh Valley 401* was fitted out with heavy-duty lifting gear such as 50-ton rings, 50- and 20-ton blocks, and 15-ton turnbuckles; it was used to lift railroad cars.

The *San Diego's* remaining propeller was blown off in the autumn of 1973. With two blades sticking up out of the sand, it lay along the hull for nearly eight months while its recovery was organized. In July of 1974, the *Sailfish* towed the *Lehigh Valley 401* to the site. Using a 5-inch diameter polypropylene hawser, the barge was anchored onto the shaft from which the propeller had been separated.

The *Lehigh Valley 401*. (Courtesy of Ed Betts.)

Ed Betts dived and supervised all underwater work. His original intention had been to lift the propeller with two steel tanker moorings he had adapted for salvage; each pontoon weighed 5,000 pounds, had twenty tons of lifting capacity, and could withstand an internal pressure of two hundred pounds per square inch (thus obviating the need for over-relief valves.) Once the propeller was floated, the tug could tow it to a dock where it could be hoisted onto a flatbed truck for overland transport to the scrap yard with whom purchase arrangements had already been made.

George Dyott wanted to have the barge's boom laid flat on its hull; using the length of the barge as a lever, the propeller could be hoisted to a point just below the surface, and secured. Then, the *Sailfish* could tow the barge and the attached propeller to the dock.

Frank Davidson wanted to lift the propeller straight out of the water, lay it on the deck of the barge, then hoist up additional bronze pieces such as the propeller shafts. Despite the protests of Betts and Dyott, Davidson's plan was adopted.

The boom was positioned thirty degrss off the starboard side of the barge. A cable that was run through the block on the derrick was connected to the propeller. After the boom took up the strain, Betts dived to make sure that the cable was clear of the wreck. When he surfaced and gave the okay, Davidson eased the lifting engine into low gear; slowly, the cable was rolled onto the rotating drum.

After several minutes, when Davidson felt the propeller was clear of the wreck, he put the engine into high gear. Unknown to him, as the

Salvage equipment and dive gear on the deck of the *Lehigh Valley 401*. (Courtesty of Ed Betts.)

propeller rose past the towering hull, it caught under the end of its own shaft and momentarily wedged itself. The lifting engine pulled so hard against the unmoving propeller that it overstressed the derrick's A-frame; the starboard leg took all the strain.

A loud, resounding crack was heard by everyone, even those aboard the *Sailfish*. The A-frame's starboard leg had dropped four inches. Since the leg was connected to the outer hull by double rows of bolts, when the leg dropped it ripped four-inch-long holes into some fifteen planks, all of which were below the water line. Davidson let go the brake and put the propeller back on the bottom.

In anticipation of placing some seventeen tons additional weight on the barge's bow, the result of which would bring dry seams underwater where they would weep, two 4-inch pumps were fitted to suction lines below deck. The support crew jumped for the pumps, but water flooded in so fast that there was no time to start them.

The barge listed dramatically. Topside personnel were taken off the stern by the *Sailfish* as Dyott kept her bow hard against the upturning hull. Betts leaped off the barge's port bow as it rose high in the air. Two men jumped into the water on the starboard side. All the loose gear—air bottles, compressors, scuba equipment, generators, pumps, and a shark cage—crashed into the sea behind them. Moments later, the derrick struck the water; miraculously, the steel frame and boom missed by inches the two men treading water.

Left; The upside down *Lehigh Valley 401*. (Courtesy of Ed Betts.) Right: A week later the bow of the barge was still awash.

The barge then capsized. As the derrick slipped under the water, the men escaped entrapment and were picked up by the *Sailfish*. With the bottom exposed to the sun, The *Lehigh Valley 401* slowly settled by the stern, drawn down by the weight of the machinery. The barge pivoted on the vertical A-frame; the bow rose as the stern submerged. In seven or eight minutes it was all over. The stern sank to the bottom; the upside down bow protruded six feet out of the water, held up by the strength of the derrick and by the buoyancy provided by the wooden hull.

No injuries occurred during the tragedy, but follow-up events were more than insult: they were piratical. The very next day, while the salvage crew was regrouping and making plans to recover their equipment, divers on a boat from Brooklyn attacked the site and took as much gear as they could. By the time the *Sailfish* returned, much of the outfit had been stolen.

The Coast Guard asked Davidson if he wanted to abandon the wreck. Davidson did not. The Coast Guard then placed a wreck buoy near the site as a warning to mariners that the exposed barge constituted a hazard to navigation. Davidson was told that as long as he retained ownership, he was responsible for the buoy and was liable for damages resulting from marine casualties, such as collisions, caused by the wreck. Under maritime law, this arrangement clearly established Davidson's ownership of the barge and all property belonging and attached to it. Salvage operations were considered interrupted, not abandoned.

During the next three months the *Sailfish* made twenty-two trips to the barge in order to recover what had been too large for the thieves to carry off. Eventually, the crane of the *Lehigh Valley 401* collapsed, and the barge fell flat against the sand. It lies there still, about seventy-five feet off the starboard stern (southwest) of the *San Diego*. The buoy was later removed.

In October, Betts and Dyott completed recovery of the equipment lost with the barge, and prepared to salvage the *San Diego's* propeller by means of the original plan. The *Sailfish* towed the two pontoons to the wreck site, and sank them in position for rigging. Cables from the sunken barge stretched across the sand, still connected to the propeller.

At this point the machinations of Bob Shourot entered the picture. He sneaked out to the wreck site with his boat, the *Black Coral*, and a crane barge. Using the tackle that Betts had already rigged, he lifted the propeller onto *his* barge and towed it away. Shourot sold the propeller before legal action could be brought against him, and before the propeller could be arrested, then absconded with the money.

Furthermore, Shourot later attempted to conceal his treachery by explaining to the Department of Commerce why he thought the Local Notice to Mariners listed two wrecks where only one existed; this action hampered the official investigation by offering a gross perversion of the truth that caused a Department of Commerce memorandum to state "Mr. Shourot says owner of scow was attempting to illegally remove a propellor when it sank and subsequently reported a second sinking to conceal his original activity."

At this point the disposition of the propeller, like many of Shourot's activities, founders in obscurity.

Betts later returned and recovered one of the pontoons for use in another salvage job. The pontoons were so large they required a slip the same as a boat, and for that reason he left the other pontoon on the bottom until he needed it. It is still there today, just forward of the starboard 6-inch gun port second from the aft.

While the *San Diego* seems to have resisted vigorously large scale salvage for profit, she has opened her hull to light salvagers whose aim is historic preservation. This is fortunate because, long after the wreck ceases to exist, at least some part of the last armored cruiser can be displayed to the generations to come.

The *San Diego's* propeller shortly after Shourot brought it ashore. (Courtesy of Steve Bielenda.)

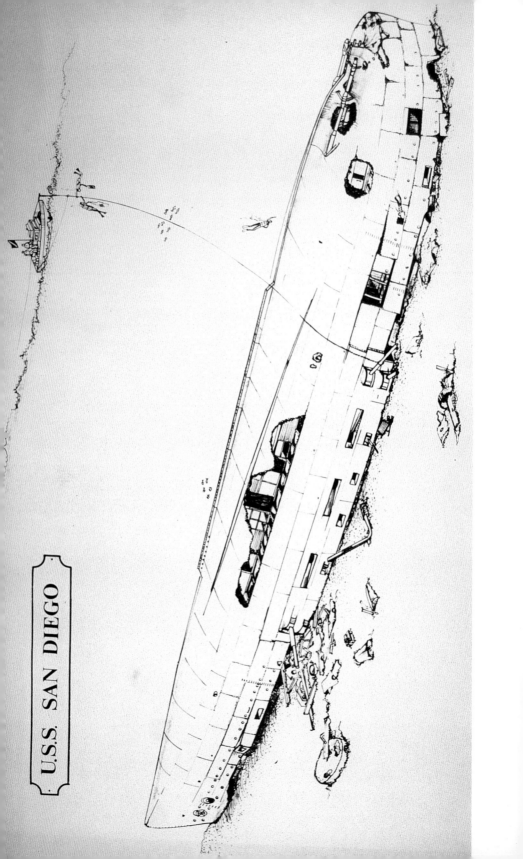

CHAPTER 6
PRESERVATION

The sea is a sacrifical element: a weak but ever-present corrosive that consumes whatever lies exposed to its soup of chemicals. Aluminum dissolves, steel and iron oxidize, copper corrodes, brass pits. Teredos, those wood-boring mollusks with a voracious appetite for wood, scythe through planking and timber will little abandon. Bacteria, unseen to the naked eye, devour all forms of organic material. Eventually, all the handiworks of man are reduced to their constituent parts: molecular debris without form or function in the human realm.

But even as the wreck of the *San Diego* collapses, the heritage contained within is being preserved. Credit goes not to the fisherman for whom the hull was saved, nor to the archaeologists who cry for historic sanction but do nothing about it. Archaeologists work only for money and profit, steadfastly refusing to shell anything out of their own pockets, relying instead on investors (grants) and unwilling backers (taxpayers who support them with their hard-earned wages via tax diversion) to support their habits. They manipulate legislation to advance their private indulgences. They claim to want only historic information, but in actuality thirst for power and possession.

This volume is written for those who care enough about the *San Diego* and the history she embraces to exert their energies toward actual exploration of her remains. To challenge the obstacles put forth by nature and greedy adversaries the sport diver ignores the complacency of his air conditioned living room and comfortable easy chair in order to see for himself that majestic shipwreck before it disappears through natural erosion and legal avarice.

The huge upside down hull appears from a distance intact and impenetrable. The steel reef is enameled with marine encrustation, pimpled with goose barnacles, carpeted with grass and short-branched seaweeds, and overrun with great frilled anemones whose tentacles wave seductively in the surge. About this microcosm flit black sea bass, tautog, ling, and the omniscient cunner, or bergall. Occasionally, schools of pelagic fish swoop down from the sunlit surface to feed upon what has become a steady, resurgent ecology.

Although the depth of the surrounding sand is one hundred feet, the *San Diego* has scoured a local excavation that is ten feet deeper: the same way waves wash sand out from under a child's dainty feet as he contemplates the sea from the surf line. The mammoth steel hermitage towers upward to within sixty-five feet of the surface, where the longitudinal

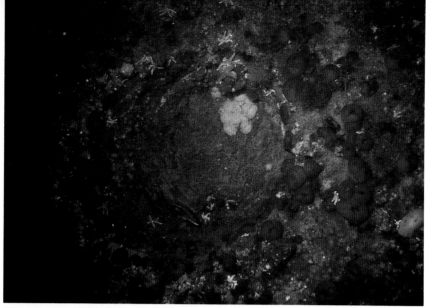
The outer door of the starboard torpedo tube.

docking keel protrudes above the hull like the ornamental backbone of a prehistoric dinosaur.

The two bilge keels are shorter than the docking keel; curiously, while the port bilge keel parallels the keelson, the starboard one is angled to such a degree that the forward end terminates within touching distance of the docking keel. The purpose of the bilge keel was to help stabilize the ship's roll in the same manner of a sailboat's deep keel.

At the bow, both anchors are snugged in the forward hawse pipes; the after pipes are empty, leftover from the days of original construction when folding anchors that were hauled up and stored on angled deck plates filled the hawseholes with massive linked chain. The torpedo tube indents are sealed by the outer tube doors.

The port propeller shaft and A-frame. (Photo by Jon Hulburt.)

At the stern, the two propeller shafts jut out of their glands and point aft like fat, horizontal columns. The starboard shaft is supported by the A-frame shaft bearing. The port shaft runs through the bearing, but A-frame dangles over an immense rust hole that exposes what used to be the after steering room; the shaft is bent by its own weight, slowly levering itself out of the structural support gland. The rudder has long since disappeared; perhaps it lies buried in the sand.

The shifting sands sometimes expose the bronze letters at the fantail. Several, but not all, have been removed.

The *San Diego* lies approximately north-south, with the bow heading north. Because the wreck is upside down, confusion arises when describing the sides as port and starboard; looking forward along the hull the diver's right hand faces the port-side of the ship, his left hand starboard—the opposite of nautical terminology. Therefore, the two sides of the wreck are designated either the "good" side and the "bad" side, or, preferably, the east side and the west side.

The "bad" side earned its epithet from the unbroken steel wall that confronts the diver swimming along the east side of the hull. Along the sand, from bow to stern, there are no openings except for the two after gunports. This is due to settling of the original uneven tilt which forced the port-side gun ports under the sand.

The "good" side contains the debris field formed by the superstructure wreckage. As the *San Diego* rolled to port, the masts, funnels, and everything loose that slid off the upper decks, fell into the sand on the starboard, or west, side of the wreck. The midship area is littered with broken beams, slices of metal, and unidentifiable iron trash.

The largest concentration of debris lies adjacent to the forward 6-inch gun barrels, which protrude like pointers into the remains of the bridge works. Brass trappings can be found here if the current scours a trough through the wreckage, and often brass glistens like gold as it is blasted clean by swirling sand. About forty feet from the hull, the circular observation platform from the forward mast lies inverted in the sand.

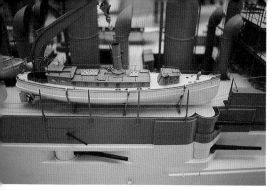 

Left: A motor boat on the model of the USS *Pennsylvania*. Right: the propeller of one of the *San Diego's* motor boats.

Slightly aft, and away from hull, lie the remains of two lifeboats that went down with the ship. They are upside down and partially buried. About midships and close to the hull one davit beckons awkwardly like a curled finger, pointing at the sky. Off the after 6-inch guns lies the crow's nest and the top of the stern mast. The cylindrical object partially buried against the hull halfway between the 6-inch guns and the fantail is a lifting pontoon from the aborted propeller salvage.

Upon closer examination the hull is found to be breached by numerous cavities: jagged explosion craters, gunports, missing hull plates, and rust holes, allowing easy access to the vast expanses and honeycombed passageways of the dark interior. Inhabiting this inner sanctum are lobsters, rock crabs, sea ravens, and ocean pouts. Few other creatures live inside unless a continuous current flushes fresh water through the open compartments.

The inside of the *San Diego* is dull and drab, colorless, and full of thick mud, suspended silt, and easily dislodged conglomerates of rust. Surprisingly, when the water outside the wreck is laden with plankton and floating, churned-up sand, limiting visibility, the water inside is often clearer because the sediment does not readily enter the structure.

During times of great surge one must exercise caution when approaching any opening in the hull. Once, while pointing my light into a porthole, my arm was sucked into the wreck with such force that my face was pinned against the hull and my mask and regulator dislodged. I was trapped for several seconds until the surge reversed and I was expectorated like a crumb coughed through a giant's teeth. In such a circumstance a

Left:. The after crow's nest. (Photo by Jon Hulburt.) Right: Be careful not to get sucked into an open porthole. (Notice the drip pan "above" the porthole.)

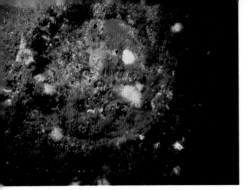 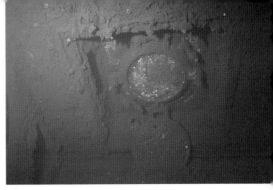

The outside of the portholes are covered with colorful marine growth, while inside they are drab and practically lifeless.

sharp protrusion of metal can easily slice through the fabric of a drysuit and gash delicate skin with expensive and harmful results.

The forwardmost point of entry is a rust hole in the west-side hull leading into a compartment below the rope locker; it is high on the hull and aft of the hawse pipes. At first glance the room appears empty, but a careful look reveals thickly encrusted ropes on a shelf overhead. Through the compartment's after hatch is a warren of narrow, debris-filled corridors leading to other tiny, clogged compartments. Fair sized rust holes in the outer hull let in enough light to dispel the gloom. This is a storage area that is fascinating to explore, and easy to get lost or stuck in. Because the diver must touch so much of the confining bulkheads and deep debris pile, often banging his tanks on the overhead and scraping his belly and weight belt over loose material, the rooms silt up quickly. The area is also complicated to comprehend due to the offset levels and inverted orientation.

It is possible to work completely across the wreck and emerge in the bread room on the opposite side, and from there exit the hull through one of the rare east-side openings. Once outside, one can swim ninety feet aft on the same level to a large hole below and in line with the forward end of the east-side bilge keel, and enter the forward boiler room. By some clever dodging and zigzagging it is possible to pass through the boiler rooms and center line bulkhead and leave the wreck through the mammoth west-side hole.

Alternatively, once inside the forward east-side boiler room one can go forward into the ammunition passage that hugs the hull of the lower platform deck. Inboard doorways lead to the 6-inch handling room, 6-inch

Left: 6-inch powder cannisters still in their racks. (Photo by Jon Hulburt.) Right: The gunpowder looks like tanned cigarette butts with holes drilled through their lengths.

Left: Jon Hulburt and all his paraphernalia enters through the gun port. Despite my careful movements, my exhaust bubbles are already dislodging rust from overhead. Right: A toilet in the crew's washroom.

shell room, and 6-inch magazine (powder room). Divers have removed many of the four-foot-long copper and wood-lath watertight tanks containing the silk powder bags filled with powder. This area is completely without light.

Any interior exploration should be undertaken with great caution and deliberation. The wise diver who plans his dive will obtain deck plans of the *San Diego* and study them closely. This will not only lead him toward the joys of self discovery, but may save his life. For a complete description of wreck penetration techniques, see this author's book, the "Advanced Wreck Diving Guide." On a personal note, in order to attune my mind to the actual condition of the wreck, I always look at profile plans upside down. Remember, too, that on the deck plans the position of the outboard rooms appear reversed because the wreck lies upside down. Also, most of the plans available were copied from original drawings; because of refits and alterations during the course of the *San Diego's* career, they may differ slightly from the way the ship appeared at the time of her sinking.

The west-side, forward 3-inch gun port is a little tight, but large enough for a diver twisting sideways to enter. It opens into a large room that can also be entered by the next 3-inch gun port aft; one can easily swim from one gun port to the next, and on days with bright sunlight and good visibility one can see both openings at the same time from a point halfway between.

Forward of this large room and just inside the forward gun port is the crew's washroom with toilets overhead. The forward bulkhead is partially rusted through, as is the inboard bulkhead. The adjacent passageway inboard of the washroom can usually be used to reach the crew's water

Left: Leather boots. Above and right: Sometimes it takes a sharp eye to detect artifacts among the encrustation.

closet forward, although sometimes the height of the sand dune prevents easy access. In that case one can squeeze through one of the washroom's rusted bulkheads.

Toward the interior of the wreck one can penetrate only about halfway, just to the other side of the 8-inch barbette, again depending upon local conditions which can change from year to year. Now the actual angle of tilt becomes apparent, for as the wreck leans downward toward the east side, the build-up of sand and silt remains level. Progress is blocked when the overhead slants down below the dune. This is unfortunate because it means that most of the *San Diego's* loose appurtenances are buried.

As the ship rolled to port after the flooding of the engine room, all items not bolted down slid across the slippery decks until they came to rest against the port side (unless they caught on longitudinal partitions that divided the interior spaces into rooms.) Later, years after coming to rest in that unnatural position, the partitions collapsed and allowed much of what had previously been captured on the starboard side to tumble or roll through gravity and the action of the sea until most of the stock and staples lay against the port hull, later to be interred under the accumulation of sand, mud, and debris. I estimate conservatively that throughout the wreck at least three-quarters of the *San Diego's* artifacts wait in this manner to be uncovered.

Halfway between the two forward gunports and slightly inboard is an overhead hatch leading to the berth deck and, appropriately, to the crew's berthing space. For some reason an aggregate of mud surrounds the hatch coaming to a depth of several feet, as if the room were filled with muck and

The pitcher shelf yields white, unmarked pitchers. The pitcher shown at far right, with the USN monogram, I found elsewhere.

the hatch had become a siphon. Ascending through this rectangular companionway one feels like a gopher climbing out of his burrow at night. Boots and tatters of cloth can be seen protruding from the slope.

Straight overhead and just inboard is a shelf that is actually a narrow storage space between the berth deck and the protective deck. Originally, this space was accessed through circular trunks sealed by manhole-type covers. In this particular spot porcelain water pitchers were stored, and the shelf is what remains of the deck as it rusts back from the outer hull. It should be understood that these storage spaces exist all the way across the ship forward and aft of the armored bulkheads that protect the engine and boiler rooms, except for the region around the turrets.

Once in the berth deck the entrance hatch becomes extremely difficult to relocate. Descending silt and rust obscure visibility and, since it falls in the direction of escape, can make for some scary moments. A little light peers in through open portholes, but not always enough to right disorientation.

One can swim aft along the west-side hull, past a bathtub, and enter the open and cleaner area of the sick bay. About ten feet inboard a long rust hole in the deck provides an easy exit down to the gun deck which is also open and unobstructed. From here one can reverse course and go forward, and emerge from the hull via the second 3-inch gun port (which is larger than the first one and easier to fit through.) One must not panic, however, if a blank wall is encountered before reaching the gun port. This bulkhead juts inboard only eight feet. Yours truly has hatched quite a few goose bumps when feeling his way out of the wreck in limited visibility and banged his mask into the steel barrier.

Left: Operating table. Right: Autoclave and steam reservoirs.

Left: Glass tubes of catgut freed from the mud float up and collect overhead. (Photo by Jon Hulburt.) Right: Dispensary bins.

Alternatively, one can explore the sick bay aft where it merges with the dispensary. The big cylinders standing in this room are the steam reservoirs for the autoclaves, or sterilizers, one of which I removed and recovered years ago. Wooden cabinets protruding partially out of the mud yield pill bottles and assorted medical supplies.

Inboard, one can work through the remains of the refrigerator room, band room, and paymaster's issuing room, and reach the east-side passage. This is because much of the deck has rusted through, allowing the mud to sift through the holes into the gun deck. The two companionways just forward of the armored bulkhead below are completely gone. One of the hatches overhead leading into the 8-inch handling room can be forced, but the blackness in there is absolute.

One can also penetrate the armored bulkhead through the doorways that would have aligned with the now nonexistant corridors. Beyond this point, in the middle of the ship, is an archway that surrounds the base of the mast. By continuing along either corridor aft one enters the laundry room. Since there are no portholes here there is no comforting glow to aid orientation, and there are no other exits unless one goes all the way through the ship and out the opposite transverse armored bulkhead, or finds a rust hole down to the gun deck. It is a spooky area.

Dropping down from the dispensary just forward of the armored bulkhead, and onto the gun deck, another set of doorways penetrate the armored bulkhead. This is another nether region that is dark and dismal, and, because of the sagging berth deck, is not as high or as open.

Left: The autoclave's pressure gauge. Right: A museum display of dispensary items—in the background is the autoclave, in the left foreground is the bottle rack shown on the next page.

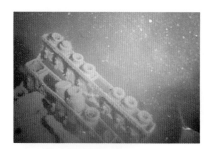

Left: A boxed lunch. Right: Rifle butts and stocks.

One can continue the descent from just forward of the armored bulkhead and exit onto the main deck, in essense leaving the hull of the wreck by dropping into the broad washout underneath. Here one can see the wreckage that has fallen from above, and easily swim out from under the side into the debris field, or explore forward as far as the forward 8-inch turret. Watch for items that have fallen through the decks that may not have yet been covered by sand.

Heading aft one passes the two 6-inch gun barrels, whose breeches and sealed port have effectively blocked entrance into those compartments. The lower gun, within the casemate of the main deck, is sometimes buried by sand. However, one can enter the casemate through any of the next four 3-inch gun ports in the midship section, despite the fact that the forward-most and aftermost have the guns in them. These ports are roomy. Oddly, although the casemate is below the level of the sand, the prevailing current sweeping through the wreck usually keeps the main deck inboard relatively clean.

Even roomier, however, are the three central 6-inch gun ports on the gun deck. Since these guns had been removed prior to the *San Diego's* sinking, they are almost like garage doors in relation to the size of a fully geared diver. The vast room inside no longer has standing partitions, allowing one to swim into the first gun port, around the inward-jutting bulkhead and past the second gun port, then around another inward-jutting bulkhead and out the third gun port. By extension, one can also reach the forward and after 6-inch gun stations which still have their guns, although here one is penetrating to an area without light; furthermore, the most

Rifles still locked in the racks.

 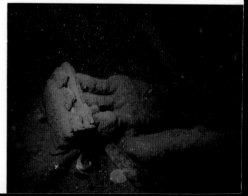

adventurous can go all the way past those guns and out through the armored bulkhead doors. Think about it before you try it.

Inboard of the central 6-inch port, about midships, is the armory. Most of it is buried under mud and silt, but racks of rifles can still be seen protruding from the muck. Unfortunately, all the iron parts of these Springfield rifles have rusted away, leaving only the carved wooden stocks.

Access through these large gun ports is so easy, and the space inside is so large, that the unwary divers can be lulled into a false sense of security. The steel plate is rusted through between the gun deck, where one enters the wreck, and the berth deck overhead. Because of the downward tilt of the decking one can swim eastward on the gun deck, add buoyancy in order to rise above the stirred up silt, and return westward only to encounter the solid outer hull with no ports and not a glimmer of light. The disorientation occurring to such a diver at this point is absolute because he knows he just came in that way, and cannot fathom that he has unwittingly changed deck levels.

This happened to two divers once while I was exploring the junior officers quarters aft of the armored bulkhead. Neither one of them knew where they were, how they had gotten there, nor how to extricate themselves from their predicament. They swam back and forth seeking a way out of the impossible situation. From more than fifty feet away I happened to be passing the west-side after armored bulkhead door when I saw a feeble light swinging in an arc where I did not think anyone other than Jon Hulburt had a right to be. I pointed my light along the corridor and slowly waved the beam, figuring Jon would ignore me, but anyone else might like to know that someone was nearby.

Instantly the light stabbed straight in my direction, and raced forward like the headlight of a speeding freight train. I was nearly bowled over as the lead diver crashed through the door slashing his index finger across his throat. I offered him my regulator, but he declined it, as did the other panic-stricken diver tumbling out of the dark passageway. Instead, I led them out of the wreck by the shortest route possible, and took them to the anchor line. They were not completely out of air, but pretty low—low enough to scare the hell out of them.

Their gratitude was overwhelming. Their thanks began by offering to buy me dinner when we made port, grew to replacing all my dive gear, and wound up with them proposing to buy a new Chevy Blazer for me. This time, it was my turn to decline. I was just happy they were alive—but not as happy as they were.

Above the three empty 6-inch gun ports extends a massive rip below the armor belt (above, if you look at the wreck upside down) that affords access to the west-side boiler rooms. Only explosive demolition can account for the hundred-foot wide jagged scar; I suspect that the mine reported by the Merritt & Chapman divers eventually detonated, not only peeling back hull plates, but initiating the process of decay by further loosening beams

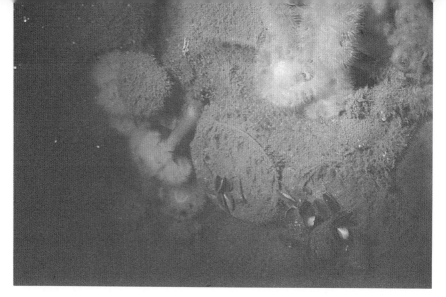

and rivets. Once this region of the wreck was weakened structurally, its further collapse was accelerated. Since I first dived the *San Diego* in 1973 I have seen this rusty opening more than double in size.

The coal bunkers that lined the perimeter of the hull adjacent to the boiler rooms are gone, the coal having fallen to the lower decks. Passing through one of the many openings into the boiler rooms proper is relatively easy. Lots of light penetrates the outboard side of the boilers; a diver can usually look over his shoulder and see the green glow of daylight and an unobstructed path of egress.

The boilers are still bolted to their bedplates, hanging threateningly, while around them is suspended a cat's cradle of broken piping, disjointed grates, and collapsing catwalks. The fire brick went the way of the coal. As one slips between the naked boilers one is confronted with increasing darkness; separation from a quick and easy exit becomes more pronounced.

Even without exploring farther afield one can unknowingly become lost in the maze. The bulkheads between the boiler rooms and fire rooms have

fallen away, so that one can traverse this immense interior space unopposed. Furthermore, the room is so tall that it is deceptively simple to return to one's point of entry and not see it because of an unperceived change in depth: one might pass inches above or below the hole that was previously so obvious.

The longitudinal bulkhead separating the east-side boiler rooms from the west-side boiler rooms is still intact. The open doors beckon like the Sirens of Argonaut fame, but the darkness beyond should act as a deterrent to all but the most prepared. The east-side boiler rooms are a replica of the west-side rooms, with abysmal lack of light added for discomfort. With a great deal of time one can eventually work his way out a hole in the hull forward of the forward boiler room, but do not count on it. It is also possible to reach the engine rooms by heading aft, but the route is circuitous and requires great resolve to navigate. Progressive penetration, as always, is the way to salvation.

The engine rooms are more easily accessible from the east-side mine crater: a ten-foot-high, twenty-foot-wide hole large enough to drive a tractor-trailer through. Both engines are suspended from their bedplates, but tons of auxiliary equipment have either fallen or are in the act of falling. One can work forward through the debris and reach the boiler room area, or cross the longitudinal bulkhead and enter the starboard engine room. It is possible to locate a doorway on the west side which opens into the cofferdam, and from there go either forward and out a rust hole in the hull, or aft into the ammunition passage; however, because of the lack of light the cofferdam is not easy to locate since the opening looks like just another corner of the engine room. The reverse course is preferable: entering the cofferdam from outside, or from the end of the ammunition passage, and using it to find the starboard engine room door.

The east-side cofferdam, adjacent to the mine crater, has been obliterated, but one can swim aft along the inside of the hull and go through the doorway into the east-side ammunition passage (on the lower platform deck), and from here go through any one of three consecutively reached inboard-facing doors into either the 6-inch magazine, the 6-inch shell room, or the storage room adjoining the small arms room. By not entering these doors, and by continuing aft, the ammunition passage leads to the after steerage compartment.

The after portion of the *San Diego* is a complicated, three dimensional labyrinth which is fascinating to explore both for those adventurous in spirit and for those seeking souvenirs of their sojourn. Numerous yawning cavities punctuate the hull, giving it the appearance of a multistory office building still under construction. The potential for exploration and discovery is unlimited. Because all the interior spaces are somehow interconnected, and because many of the rooms open into adjacent rooms each branching again into other rooms, I can give at best a restricted overview of this complex warren.

Looking aft from the 6-inch guns, entry is made by crawling under the wreck. (Photo by Jon Hulburt.)

Starting at the bottom and working up, the wreck can be accessed by swimming under the hull about ten feet aft of the after 6-inch gun barrels. Unlike the bow of the wreck, here the turret is not visible; only a narrow crawl space exists under the main deck, and then for only a short distance. By angling forward one can slither through the wreckage of the engine room skylight and into the engine room, and from there upward. By angling aft along a trough one can curl up through what used to be a stairwell and emerge in the gun deck within touching distance of the west-side door in the armored bulkhead.

One can easily go through the armored bulkhead door, but will find that the settling of the hull has forced the superstructure to buckle up through the engine room skylight, offering slender passage. Heading forward one can reach the after 6-inch gun. Rising, one can continue through the skylight rubble and reach the berth deck level, and then go out the berth deck armored bulkhead door and into the berth deck proper. One can also turn east, follow the inside of the armor, and either tumble out the east-side armored bulkhead door, or reach the dark side of the engine room compartment. This is a black, mud-filled area.

Left: Skylight porthole. Right: Sheets of linoleum.

Left: Speaking tubes. (Photo by Jon Hulburt.) Middle: Control throttles. Right: Telephone box. (The wire hoops "under" the brass box contained glass "bombs" fill with carbon tetrachloride, which could be hurled at the base of a fire.)

The gun deck aft of the armored bulkhead is officer's country. West of the skylight is the paymaster's office and the navigator's office; farther aft along the hull there used to be four ward officer's rooms and the Chief of Staff's bath. No partitions remain standing, although one can see pieces of furniture, exploded desks, and the bath tub, where the rooms used to be.

Inboard lies a skylight trunk, and on the other side more offices. One can work aft along the east-side past the 8-inch barbette without too much difficulty, and on a good day one can still see light coming in from the west-side portholes. It is here that one becomes woefully aware of the *San Diego's* extreme tilt, for no portholes are exposed on the east-side (either inside the wreck or outside) and the mud has accumulated right to the overhead.

The captain's pantry used to stand just aft of the barbette, but is buried under the sand. However, one can ascend through the skylight trunk into the berth deck and into the midst of the ward officer's and junior officer's pantry, where china can still be found. Alternately, one can ascend to the berth deck through the larger ward room skylight trunk further aft.

Staying on the gun deck one now encounters a partial transverse bulkhead, the middle of which is still in place as are the two sections protruding inboard from either side of the hull. Beyond this bulkhead one

Left: Exploded desk. Right: Dishes in the pantry.

enters a room that has empty 3-inch gun ports on either side. Sometimes a sand dune exists in the wreck's centerline that nearly touches the overhead, making a cross passage impossible; usually, one can mount the dune and see both gun ports.

Here again the wreck's angle of lean becomes apparent. The rim of the west-side gun port is over six feet above the level bottom, while the east-side gun port gouges *below* the sand. Yet a severe current rips through the east-side gun port often enough to keep it dug out to a depth of 120 feet: the huge crater is clearly evident as one climbs up the steeply sloped walls and realizes, looking forward, that not only are the gun deck portholes buried under the natural level of the sand, but so are the berth deck portholes ten feet higer on the wreck.

Because of the constant scouring effect, the deck inboard of the east-side gun port is kept clean. Often, loose items slipping out of the central sand dune roll down the hill and lodge against the gun port's rim. The surge can be terrific. Once, as I approached the edge of the gun port, I was sucked out by the incredible power of the sea, spun around two or three times, and spat out at the top of the hole outside the wreck. After I regained my senses I had to time my reentry with the incoming surge so I could recover my goody bag left inside.

The space between the two gun ports is the captain's saloon: not a place where the captain got drunk at sea, but a lounge. His quarters are buried deep under the mud just forward of the east-side bulkhead protrusion.

 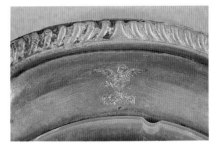

Everthing aft of here is admiral's territory, occupied only when the *San Diego* was designated flagship of the fleet. Unfortunately, except for three or four feet of space, most of this area is under sand and debris that fell from the decks overhead.

This does not mean that the admiral's quarters are impenetrable, for one can duck under the drooping beams without difficulty, especially since the overhead separating the gun deck from the berth deck has long since collapsed. In fact, both decks merge from this point aft into one huge room, making it difficult to discern which deck level one is probing. On a clear day light streams in through double rows of portholes lining the sides of the wreck. Both after 3-inch gun ports have toppled out of the hull, and the holes have rusted back so far that the openings are no longer recognizable as rectangular gun ports. Hull plates have sheared off the berth deck beams, leaving gaping wounds, and other plates are loose and in the process of collapsing outward. The naked, arched support beams are reminiscent of a flenced whale whose mammoth ribcage lies bleaching in the sun.

By rising slightly and reversing course one can swim forward until a point is reached where decking reappears, and where one can definitely concede that one has gained the berth deck. At the same place the east-side gun deck portholes dip beneath the sand. Ample light guides the diver through an abnormally high room made taller by the missing protected deck, above which storage areas were once sequestered. Before reaching the frame where the next-forward pair of 3-inch gun ports pierce the gun deck below, the inside of the east-side berth deck portholes become covered with a hill of mud. Outside, however, one can still see glass; I have often thought of punching out the panes in order to permit the mud to siphon through the ports, but have been deterred because the storm covers appear to be battened down.

At this point what appears to be the east-side hull of the berth deck is in reality the storage area of the protected deck, The actual berth deck is completely buried. The center part of a transverse bulkhead blocks passage through the middle of the wreck, but this steel barrier can be bypassed easily on either side. Just forward is a skylight which one can drop through to gain the gun deck. To the west is a three-foot-square hatch which a diver can use to leave (and enter) the berth deck.

Left: A bronze hatch to the protected deck. Right: Stores.

In the 1970s the square hatch and the skylight were the only points of entry into the berth deck, and at that time most of the studs and many of the partitions still stood. Whereas now the berth deck from the armored bulkhead aft is one gigantic, open room extending all the way to the fantail, a distance of some 150 feet, then it was sectioned off like the interior of a partially completed house. And whereas now much of the silt has been swept clean by the continuous current, then it lay like a pall and was quickly agitated.

During one excursion into the innermost recesses I stumbled into a dark room with three walls, kicked up enough silt to reduce visibility to

Upper left: The so-called garage door opening, on the west-side berth deck just aft of the armored bulkhead. (Photo by Jon Hulburt.) Upper right: An indicating panel which appears to be identical with one on the USS *Olympia* at lower left. Remember that the panel on the *San Diego* is upside down. Lower right: A similar panel.

The central hub and A-frame shown left is one of three that hangs upside down in the after steerage room, all that remains of what would have looked like the auxiliary steering wheels on the USS *Olympia*.

nothing, and lost orientation. In a last ditch attempt to find my way out, after going around in circles for several minutes, I took a compass bearing. Knowing that the wreck lay north-south, I swam west. Although the mass of surrounding steel precluded pinpoint accuracy, the needle was steady enough to give me an approximation of course. I bumped into the west-side bulkhead and followed it until I knew it eventually had to lead me to an exit. It did, and I lived to tell about it.

On the boat I was white as a ghost from the experience, while my dive buddy, Carol Coleman, was calm. My fear was evident as I explained the unorthodox method I used to get us out of the wreck. Carol, who stayed with me throughout the maneuver and followed me blindly, trusting my experience, had no idea we were lost. Ignorance, indeed, is bliss.

No longer do the officer's staterooms stand along the hull. The west side is an empty expanse where entire hull plates weighing several tons have sloughed off (called the garage door), while the east side is clogged with mud that covers the portholes. The pantry is nonexistent, the only structure still standing being the 8-inch barbette and some surrounding beams. One can swim on the east side of the barbette, through the armored bulkhead door, then, either forward along the dark side of the engine room into the nether reaches where none but Jon Hulburt dares to tread, or west in front of the engineer's log room to the west-side armored bulkhead door. From there the brave can go forward along the corridor between the skeletal remains of the engine room skylight and the lavatory, or turn aft and go through the west-side armored bulkhead door back into the junior officer's area.

Going outside, and skipping up to the large hole just below the propeller shaft gland, one faces the entrance to the small arms room. One can go aft along the ammunition passage into the open space of what used to be the after steerage compartment, passing along the way an inboard storage room, or forward along the ammunition passage, past three inboard doors two of which lead into the 6-inch powder room, the third of which leads into the 6-inch handling room, before passing the large opening into the engine room. Farther forward one can exit the wreck through a large rust hole in the outer hull.

Entering the small arms room one is immediately confronted with boxes of rifle ammunition in five-bullet clips, single shot .45 caliber pistol bullets, and wooden cases containing starter cartridges for the deck guns. One can turn aft, go down a debris field and into a storage room, then out through a door into the after steerage compartment.

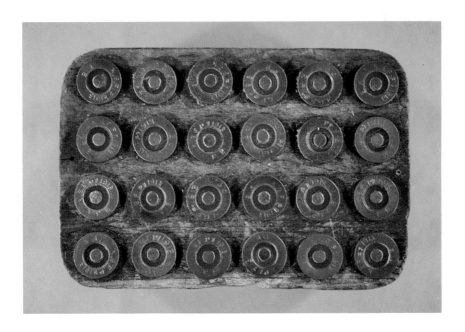

By rising to the top of the room one can head east into a storage compartment, then turn aft through a door into another storage compartment, go down through yet another storage compartment full of copper tubing, and be in the same compartment one previously reached just forward of the after steerage.

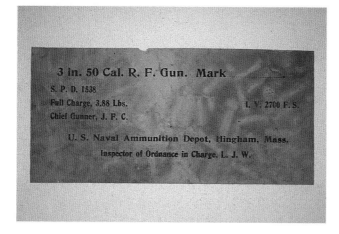

Top: 3-inch shells stacked four in a crate, and alternating nose to tail. Middle: A gunpowder loading slip found inside a 3-inch gun shell. Bottom: The ends of two saluting shells, with the primers removed.

Just forward of the small arms room is a hole leading down into the 8-inch magazine. Level with the small arms room one can go forward into the 6-inch shell room, where the projectiles lie scattered about, and both the port and starboard 6-inch powder rooms. Divers have removed most of the powder cannisters from these magazines. One can go completely through the west-side magazine into the 6-inch handling room, which has two hatches leading up and two down.

Upward is the 3-inch magazine, where most of the ammunition is wedged against the after bulkhead. Downward is the blower room, which has two trunks leading down into the berth deck near the garage door opening. On both the east side and west side of the blower room are compartments where the saluting shells are stored. In the middle of the blower room facing aft is the 8-inch handling room. One can enter this and access both the port and starboard 8-inch powder rooms. The 8-inch shell room is at the after end of the handling room. All of these rooms are totally devoid of light.

This descriptive tour of the *San Diego* is necessarily brief, and obviously impermanent. In any shipwreck the destructive process is dynamic. Lately, the *San Diego's* deterioration has been accelerating at an alarming, exponential rate. I used to peer up at the west-side hull in the berth deck and see what looked like a star-studded sky: the rivets had fallen out of their slots, and the green glow of light shone in through the holes. Now entire hull plates fall like leaves in autumn, and are quickly buried by the ever-moving sand.

Left: Ammunition hoist, lined with wood in order to preclude sparking. right: I found this deadeye outside the wreck; it was probably part of a lifeboat's rigging.

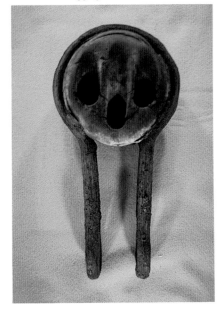

The *San Diego* is like a puzzle whose pieces keep changing shape. Jon "The Mole" Hulburt once put all those pieces together into a perceptible pattern, and made a complete transit of the wreck from bow to stern without once exiting the hull: a nearly five hundred foot penetration. This was no daredevil stunt, but a carefully planned transnavigation. Jon spent years systematically reconnoitering section by section the various rooms, corridors, hatches, and companionways, that formed his ultimate route; spent some two hundred hours studying deck plans; and kept voluminous notes on all his dives and discoveries. He avoided obstructions by changing deck levels when necessary, and made most of the midship traverse on the east side of the boiler room skylights—an infiltration he calls the dark side passage. His methodical approach and precise execution will probably never be duplicated.

The interior of the *San Diego* is no longer recognizable as a ship. Partitions have collapsed, bulkheads have been stripped away, decks have rusted through. Debris has piled up until it looks like a giant junkyard full of broken and obscure parts.

Amid this disarray of decayed wreckage, of silt and swirling sand, of black, unctuous mud, of twisted and chaotic wires, cables, and conduits, of rotting steel, and in the realm of utter darkness, lie the remnants of an entire ship's stores, those now-priceless mementoes of Naval architecture, and the vestiges of a hastily abandoning crew's possessions. The *San Diego* is a vast underwater archive that is being demolished along with its consignment of antiques.

Who is to be the custodian of these treasures? Who is to see to their safety? Certainly not those land-locked individuals who have no access to the ocean, and no interest in its products, but those who care enough to unravel the mysteries of the past by finding, exploring, and studying shipwrecks. With painstaking effort the sport diver is the unpaid historian, the diligent researcher, and the preserver of this dwindling supply of relics. While many may talk about saving our underwater heritage, it is ultimately the wreck diver who actively engages in doing it—and at his own expense.

As a shipwreck disintegrates it passes before our eyes like a movie in extreme slow motion. Each frame is ephemeral, existing for only a brief instant in time, and must be studied before it dissolves. If an artifact presents itself it must be grasped. For unless these keepsakes are removed as they appear, they will be irretrievably forfeited.

Every rust hole that opens becomes a siphon to oblivion. Every beam that collapses crushes valuable glassware. Every bulkhead that falls down buries articles of brass, silver, and gold. Whenever someone is not there during the exact moment that an artifact is visible in the debris, it is swept away.

To ignore that these processes of deterioration exist is to supplant unfounded beliefs for stubborn fact, emotion for truth and reason, fantasy for reality. The *San Diego* is but an example of the thousands of shipwrecks that are being devoured by the hungry sea. The ocean is a great leveler.

Preservation does not mean leaving an object in a hostile environment. It means removing it to a place of safety, now, while it still exists, while it is in one piece. Then and only then is this rich trove, this heritage from the past, actually preserved for future generations to see and enjoy.

Attend any underwater conference, go to any dive club meeting, visit any wreck diver's home, and see how history is *really* preserved. Here are the people who are willing to spend their own time and energy recovering, restoring, and displaying the testimonies of yesteryear. Here are the people who give legacy new light, new appreciation. Here are the people who not only preserve artifacts, but cherish them.

Wreck divers are a special brand of people. We owe them a great debt of gratitude for recovering our investment in the past—before that investment is lost forever.

California. (Courtesy of the Naval Historical Center.)

San Diego. (Courtesy of the Naval Historical Center.)

San Diego Statistics

Built by Union Iron Works, San Francisco, California
Cost of construction (not including armament): $3.8 million
13,680 tons normal displacement
15,000 tons full load displacement
503'11" overall length
69'10.5" extreme beam
24' normal draft
26.6' full load draft
Speed (maximum): 22 knots
Machinery: 2 4-cylinder triple expansion steam engines
          16 Babcock & Wilcox water-tube marine boilers
Coal capacity (normal): 2,000 tons
Armament: 4 8-inch, 40 cal.
         14 6-inch, 50 cal.
         18 3-inch.
         12 3-pdr.
         2 1-pdr.
         2 18-inch submerged torpedo tubes
Sunk by mine laid by *U-156*
Depth: 110 feet
Heading of wreck: north
Location: 40-32-25N   73-02-30W
         26543.3   43692.9

Disposition of Armored Cruisers

    ACR-1, None
    ACR-2, *New York* (later *Rochester*), scuttled at Cavite, December 1941.
    ACR-3, *Brooklyn*, scrapped 1921.

Pennsylvania class:
    ACR-4, *Pennsylvania* (later *Pittsburgh*), scrapped 1931.
    ACR-5, *West Virginia* (later *Huntington*), scrapped 1930.
    ACR-6, *California* (later *San Diego*), sunk off Long Island, July 19, 1918.
    ACR-7, *Colorado* (later *Pueblo*), scrapped 1930.
    ACR-8, *Maryland* (later *Frederick*), scrapped 1930.
    ACR-9, *South Dakota* (later *Huron*), scrapped 1930.

Tennessee class:
    ACR-10, *Tennessee* (later *Memphis*), wrecked at Santo Domingo, August 1916.
    ACR-11, *Washington* (later *Seattle*), scrapped 1946.
    ACR-12, *North Carolina* (later *Charlotte*), scrapped 1930.
    ACR-13, *Montana* (later *Missoula*), scrapped 1930.

San Diego Timetable

1899 March 3: authorized by Congress.
1901 January 10: building contract signed with Union Iron Works.
1902 May 7: keel laid.
1904 April 28: hull launched.
1906 January 30 and February 8: dock trials.
1906 October 4 to November 12: preliminary trials.
1907 July 20: preliminary acceptance.
1907 August 1: commissioned into the United States Navy.
1908 January 18: final trial.
1908 May 14: final acceptance.
1908 September 16: first crossing of the equator.
1911 January 17: first designated flagship of the Pacific Fleet.
1911 November 22: first large ship to enter Pearl Harbor.
1914 September 1: name changed to *San Diego*.
1915 January 21: boiler explosion killed nine men.
1915 Won the Spokane Trophy for gunnery excellence.
1915 October 13: movie filmed on board.
1917 February 12 to April 7: reserve status.
1917 July 29: entered the Atlantic Ocean.
1917 August 19: Captain Harley H. Christy assumed command.
1917 November: docked at La Croisic, France.
1918 July 19: sank after striking a mine laid by the *U-156*.
1918 August 26: stricken from the Navy Register.
1919 October 21: obstruction buoy discontinued.
1921 May 16: Saliger Ship Salvage Corporation planned salvage.
1957 October 29: Maxter Metals Corporation bought wreck from Navy.
1963 March: American Littoral Society formed Underwater Preserves Committee.
1963 August: Maxter Metals salvage contract expired; *San Diego* unofficially sanctioned as a fish haven.

San Diego Deck Plans

Records of the Bureau of Ships are held in the National Archives, Washington, DC, 20408. Plans can be viewed in the reading room or purchased by mail. Check for availability and current price before ordering.

The following plans are dated 1901:

90-7-2	Outboard profile, awning, and rigging plan
90-7-4	Main Deck
90-7-5A	Cross section (aft)
90-7-5B	Cross section (forward)
90-7-6	General arrangement
90-7-7	Bridge deck and boat stowage
90-7-8	Inboard profile
90-7-9	Gun deck
90-7-11	Protected deck
90-7-12	Midship section
90-7-13	Berth deck
90-7-25	Sheer, half breadth and body
90-7-31	Body plan
90-7-32	Officers' quarters and furniture arrangement on gun deck.
90-8-7	Hold and magazines
90-8-8A	Upper platform
90-8-8B	Lower platform

The following plans are alterations dated 1917:

18531	Bridges
18535	Bridge deck and boat stowage

Bibliography

California, later *San Diego*: Deck Logs 1907–1918, National Archives, Washington, D.C.

Clark, William Bell, *When the U-boats Came to America*, Little, Brown, and Company, 1929.

Journal of the American Society of Naval Engineers, February 1907, "U.S.S. *California*. Description and Official Trials," by G.W. Donforth, Lieutenant, U.S. Navy, Member.

Musicant, Ivan, *U.S. Armored Cruisers: a Design and Operational History*, Naval Institute Press, 1985.

Naval History Division, *U-Boat Operations in the Western Atlantic During World War I*, 1973, unpublished manuscript by Richard A. von Doenhoff and Harry E. Rilley, Naval Historical Center, Washington, DC.

Navy Department, *German Submarine Activities on the Atlantic Coast of the United States and Canada*, Publication Number 1, 1920, GPO.

New York Times, July 20, 1918–July 17, 1919.

"Record of Proceedings of a Board of Investigations Convened on Board the U.S.S. *San Diego* by order of The Commander-in-Chief, U.S. Pacific Fleet, to inquire into Re ship's force, U.S.S. *San Diego* being able to repair boilers damaged in explosion, January 21, 1915, passage La Paz, L.C., Mexico, to Guaymas, Sonora, Mexico. January 25, 1915," National Archives, Washington, DC.

"Record of Proceedings of a Board of Investigations Convened on Board the U.S.S. *San Diego* by order of The Commander-in-Chief, U.S. Pacific Fleet, to inquire into The Blowing Out of Tubes and Certain Damage Done on Full Power Trial, U.S.S. *San Diego*, January 21, 1915, passage, La Paz, L.C., Mexico to Guaymas, Sonora, Mexico. January 21, 1915," National Archives, Washington, DC.

"Record of Proceedings of a Court of Inquiry Convened on Board the U.S.S. *Maui* by order of Commander Cruiser and Transport Force to inquire into the Circumstances Concerning the Loss of the U.S.S. *San Diego*, July 19, 1918," National Archives, Washington, DC.

Tzimoulis, Paul: personal correspondence, 1963.

Underwater Naturalist, American Littoral Society, various issues, 1963.

Upchurch, Otis Edward, Gunner's Mate First Class, U.S. Navy: Personal Diaries, 1914–1917.

Upchurch, Colonel Richard L., U.S. Marine Corps (Retired), "Life in the Armored Cruisers," U.S. Naval Institute *Proceedings*, January 1986.

Books by the Author

Nonfiction

Diving
Advanced Wreck Diving Guide
Andrea Doria: Dive to an Era
Wreck Diving Adventures

Nautical History
Track of the Gray Wolf
USS San Diego: the Last Armored Cruiser
The Popular Dive Guide Series:
Shipwrecks of Delmarva
Shipwrecks of New Jersey

Fiction

Underwater Adventure
The Peking Papers

Science Fiction
Return to Mars
Silent Autumn
The Time Dragons Trilogy:
A Time for Dragons
Dragons Past
No Future for Dragons

Supernatural
The Lurking

Vietnam
Lonely Conflict